In Praise of *The Next Space Ra*

"American leadership in outer space will be essential for American leadership in the 21st century. *The Next Space Race* is a sweeping review and analysis of the opportunities and challenges the United States faces in the space domain. It is a solid primer for policymakers, legislators, entrepreneurs, investors, and the public on how to ensure the United States can seize the opportunity that is outer space—providing for our continued security and opening vast new economic frontiers."

—*Mandy Vaughn, CEO and Founder, GXO Inc.*

"*The Next Space Race* highlights a grossly understated problem: a picture of where we stand relative to China and its impressive space ambitions. That makes it indispensable reading in this era of Great Power Competition."

—*Kevin Pollpeter, CNA, co-author of China's Space Narrative: Examining the Portrayal of the US-China Space Relationship in Chinese Sources and Its Implications for the United States*

"There could not be a more critical time in human history to take action on securing the resources, wealth, and overall domain of space. Harrison and Garretson provide a thorough, well-researched, and detailed analysis of the who, why, and what of space, as well as providing actionable recommendations for the United States to maintain superiority and realize significant economic benefit. This fascinating book is accessible and easy to read, making it the perfect resource for anyone from the novice to the well-informed."

—*Meagan Crawford, Co-Founder and Managing Partner, SpaceFund*

"*The Next Space Race* lays out in a holistic and systemic way what is at stake in the contest for leadership in space between the United States and China, and what America's agencies need to be enabled and directed to do. In these pages, we find a menu of concrete steps for the Executive Branch and Congress to take to secure our leadership on the ultimate high ground."

—*Richard B. Myers, General, USAF, ret.; 15th Chairman of the Joint Chiefs of Staff; President Emeritus, Kansas State University*

"Finally, legislators and policymakers have a road map for empowering our space entrepreneurs, providing for their security, and opening vast new economic opportunities for us and our allies and partners. This is a must-read for industry leaders, start-ups, and groups looking for a way to explain the value of space to their representatives."

—*Dr. George C. Nield, President, Commercial Space Technologies, LLC*

"*The Next Space Race* focuses on the growing complexity between security and economics in space and identifies key initiatives needed to ensure American leadership in space. It highlights different aspects of cooperation and competition, whether across the entire US government, in public-private partnerships, with our allies, and how to prevail in our growing competition with China. One unique aspect of this book is focusing on how to further unleash the private sector as a key element of U.S. strategy."

—Kevin O'Connell, Founder and CEO, Space Economy Rising, LLC and Former Director, Office of Space Commerce, U.S. Department of Commerce

THE NEXT SPACE RACE

A Blueprint for American Primacy

RICHARD M. HARRISON
AND PETER A. GARRETSON

Praeger Security International

An Imprint of ABC-CLIO, LLC
Santa Barbara, California • Denver, Colorado

Library of Congress Cataloging-in-Publication Data

Names: Harrison, Richard M., 1981- editor. | Garretson, Peter A., editor.
Title: The next space race : a blueprint for American primacy / [edited by]
    Richard M. Harrison and Peter A. Garretson.
Description: Santa Barbara, California : Praeger, an Imprint of ABC-CLIO,
    LLC, 2023. | Series: Praeger security international | Includes bibliographical
    references and index.
Identifiers: LCCN 2022045729 | ISBN 9781440880780 (hardcover) | ISBN
    9781440880803 (paperback) | ISBN 9781440880797 (ebook)
Subjects: LCSH: Astronautics and state—United States. | National
    security—United States. | Space race.
Classification: LCC TL789.8.U5 N485 2023 | DDC 629.40973—dc23/eng/20221129
LC record available at https://lccn.loc.gov/2022045729

ISBN: 978-1-4408-8078-0 (hardcover)
        978-1-4408-8080-3 (paperback)
        978-1-4408-8079-7 (ebook)

27 26 25 24 23   1 2 3 4 5

This book is also available as an eBook.

Praeger
An Imprint of ABC-CLIO, LLC

ABC-CLIO, LLC
147 Castilian Drive
Santa Barbara, California 93117
www.abc-clio.com

This book is printed on acid-free paper ∞

Manufactured in the United States of America

Contents

Acknowledgments

The origins of this book date back to the fall of 2019 when American Foreign Policy Council (AFPC) president Herman Pirchner Jr. asked whether it would be value added for our institution to help educate policymakers on space policy issues. We tasked one of our excellent researchers, Rebecca van Burken, to help craft an analysis of all the think tanks that were examining space policy issues, including their recent events; amount, type, and frequency of their publications; and the range of space issues covered. After careful review, we determined that it would be advantageous for AFPC to launch a Space Policy Initiative (SPI) to help America's leaders grasp the implications of the unfolding space race and inform them through a wide range of educational activities like policy papers, briefings, opinion editorials, closed-door workshops, and books.

AFPC was fortunate to have already assembled a team of strategic thinkers with access to a broad network of first-rate space experts. After several months of generating numerous publications, successful congressional briefings, and workshops, the program was picking up steam and gaining a solid reputation. Once again at the suggestion of Herman, we decided it was time to take the next step and write a book that seeks to lay out a vision for U.S. space policy.

We could not have written this book by ourselves. It is the culmination of extensive research and perspectives from dozens of interviews with space experts. Several scholars, members of the AFPC team, and our financial backer, all deserve mention.

First and foremost, we would like to thank Herman for his vision and for identifying the need to focus on space and helping us launch this project. We are indebted to our anonymous donor, who believed in us and very generously supported our research budget to make the project possible—and recognized the need to provide a broader understanding of space to policymakers and the general public.

Thankfully, we were not shaping the space policy debate from scratch, and it is important to recognize the principals who have paved the path on space policy. Rep. Jim Cooper, Rep. Mike Rogers, Speaker Newt Gingrich, and Lt. Gen. Steve Kwast, all deserve thanks for their leadership in forwarding a broader vision of spacepower for national security and prosperity.

On a topic as broad as U.S. space policy, it is nearly impossible to be an expert on everything. We turned to numerous people for guidance during the research and writing process and would like to thank Brig. Gen. Steve "Bucky" Butow, Brig. Gen. John Olson, Dr. Joel Mozer, Dr. Thomas Cooley, and Dr. Bhavya Lal for their inspiration. We would also like to thank Wayne White, Bruce Cahan, Armen Papazian, Scott Phillips, and Mir Sadat for their specific ideas on legislation, policy, and finance. Special thanks go to Chris Griffin, who helped solidify the framing questions driving our book research. Furthermore, we appreciate all the guests who generously gave their time to participate in interviews on our *Space Strategy* podcast—their insights were invaluable for the manuscript (see the appendix for the full guest list). And, of course, we express our thanks to the other book contributors: Dr. Larry M. Wortzel, Dr. Lamont Colucci, Cody Retherford, and Anthony Imperato.

Additional thanks go to our peer reviewers for the project, Dr. Namrata Goswami, George Pullen, Dr. Brent Ziarnick, and Joshua Carlson. Their guidance greatly enhanced the quality of our work. We would also like to recognize Amy Marks for expertly copy editing the book—though we should note that any errors in the manuscript are our own.

Several members of the AFPC team deserve mention as well. AFPC senior vice president Ilan Berman deserves great thanks for his guidance during the proposal writing and publication process. Our book project also received invaluable support from Mandarin-speaking researchers Kyra Gustavsen and Andrew Hartnett, along with our fact-checking researchers, Linley Himes, Margaux Miller, Dee McHardy, and Alexis Schlotterback. Lastly, over the years, the SPI has benefited from the intern research support of the "space squad" (Sydney Duckor, Thomas Falci, and Lauren Szwarc), the "space jammers" (Caillou Peña and Autumn Kearny), Alexandra Jaramillo, Joaquin Liviapoma, Melinda Madden, and Ryan Christensen.

On a personal note, Richard would like to thank his loving and understanding wife, Allyson, along with his fun-loving children, James and Nathan, for their consistent support through the course of book writing. He also expresses great appreciation for his parents, Monty and Sonia, for encouraging him to reach for the stars, and for his extended family, and the whole Ashburn Xtreme ice hockey family for their encouragement throughout the process.

Introduction

The Apollo program of the 1960s and early 1970s demonstrated American ingenuity and served as the foundation for accessing and exploring the great unknown. Unfortunately, over the past several decades, space has not carried the same prominence in America. In fact, until very recently, government space projects have languished, been given low priority, and even been partially dependent on competitors and hostile foreign powers (such as Russia).

Yet, while the U.S. government's space efforts have been flagging in recent years, those of the private sector are gathering steam. Today, the U.S. government can proudly rely on private American corporations (such as Elon Musk's SpaceX) for access to space. Corporations in the United States are developing expertise and capacity quickly—and that may help foster commercial activities and opportunities in space.

The U.S. private sector understands that space has much to offer economically. Conversely, the policymaking community has been concentrating predominantly on understanding the nature, scope, and implications of adversary military threats in space, as well as the importance of maintaining a safe space environment for the United States to conduct operations. Washington has struggled to consider the major economic benefits of developing space. Likewise, space experts have failed to articulate the great benefits of space industrialization—the manufacturing of structures in space, development of space solar power satellites, generation of nuclear power systems, and space mining—or to explain that valuable space activities cannot be achieved without focused investment and government prioritization. What is needed now is a broadening of the policy debate from one fixated almost exclusively on national security dimensions to one pursuing a peacetime strategy and considering the economics of space. Indeed, as the Blue Ribbon U.S. government State of the Space Industrial Base reports note, America lacks a North Star vision for space, at least so far.

Others, however, do not lack such a vision. While America's space efforts have been focused narrowly, the People's Republic of China (PRC) has laid out and begun to implement a sweeping national space strategy—one that could, over time, severely and adversely impact U.S. economic and military security. China's effort is driven by a singular purpose, buttressed by a state-run economy and political decision-making processes that ensure its rapid implementation. From Moon landings to plans for asteroid mining and harnessing energy in space, Beijing is beginning to exploit space to achieve its great power ambitions. And more is yet to come. China's government has laid out concrete milestones in this domain, envisioning its space efforts culminating in an Earth-Moon economic zone generating $10 trillion annually by the year 2050. Beijing, moreover, is making serious progress toward that goal.

The United States needs to structure its approach to space to ensure that it can meet or surpass PRC timelines. The pace of U.S. efforts will be driven by politics, policy, and the seriousness with which we seek to address great power competition in this emerging domain. It is time to widen the U.S. lens vis-à-vis space, from human and robot exploration to a comprehensive strategy that serves American economic, societal, and military interests. There may be some aspects of the space race that are less important—for instance, human travel to Mars would be more of a symbolic accomplishment than a strategic one. But the race for Lunar resources and solar energy collection will have massive implications for our national economic power and global standing. Whichever country can sustainably achieve its objectives first will capture the high ground in what is shaping up to be a critical strategic arena. America needs to articulate a space vision committed to a path of space economic and industrial development and to guarantee the protection of such commerce. To compete successfully against China, the United States will need to go on the strategic offensive before it is too late.

STRUCTURE OF THE BOOK

In this book, we begin by exploring the need to shift from a focus on space exploration to one of space commerce, citizen spaceflight, space mining, and space development. For long-duration operations in space, it will be imperative to devote resources to nuclear power and propulsion systems. Solar energy will be indispensable for further development in space, as well as to provide power to Earth.

Next, we examine China's plan to become a leading space power by 2045. China's space road map envisions the development of techniques for asteroid mining, the creation of nuclear-powered shuttles for space exploration, and the industrialization of the Moon to fabricate satellites that can harness energy in space. Gaining an understanding of China's vision will allow the United States to effectively compete in the space domain.

We then focus on the challenges to U.S. space security. Hostile nations are increasingly treating space as a war-fighting domain and pursuing strategic military activities that challenge American space security. We discuss natural threats, anti-satellite weapons, and kinetic and nonkinetic threats, as well as outline the contemporary and future threats from China, Russia, and lesser spacefaring adversaries.

After outlining the challenges, we consider whether American space primacy is in decline. Failing to understand and identify the technological drivers of the coming space economy, coupled with the increasing adversarial competition in the space domain, could be perilous. We detail the myriad reasons to expand the U.S. presence in space and provide historical analogues that further encourage space development.

The new strategic challenges posed by space have made it necessary to alter both military organization and doctrine. By guaranteeing the security of American space assets, the U.S. government can further incentivize commercial investment in space. This requires the United States to have a robust security presence in that domain, as well as a clear understanding of authorities and priorities for the new Space Force and Space Command and how both can complement the other U.S. military services. Here we will delve into the origins of America's newest military branch, as well as its future.

In addition to protecting military assets with hard power, soft power methods are also necessary to advance U.S. interests in space. Today, global norms in space (on issues such as the weaponization of space) remain fluid and largely unformed. Here we discuss space norms and partnerships. We explain why, as the United States positions itself as a space leader, it should foster relationships with allied nations and increase the pace of progress toward common objectives and standards in the space domain.

Next, we chart the dimensions of U.S.-China competition as viewed through the lens of the six centers of gravity in space power competition (space policy and finance tools, space information services, space transportation and logistics, human presence in space, power for space systems, and space manufacturing and resource extraction). Several points of elaboration will prove beneficial for policymakers, including (a) how to prioritize space activities across the six space sectors, especially given the finite limits on available resources; (b) understanding the degree to which the United States benefits economically or strategically from specific efforts within each space sector; and (c) understanding how the United States

compares to China in each sector. We evaluate these space sectors to identify the most promising avenues for the United States to gain a competitive
advantage over its strategic competitors and potential adversaries.

Finally, we define an American space agenda to form a unified front on
space policy. To be successful, private-sector space companies, NASA, the
Department of Defense, Congress, and the White House, among others,
must all work together to unlock the limitless potential of space. The final
chapter explores and explains this imperative and recommends actions for
Congress, the executive branch, and various agencies in the U.S. government to advance a coherent space vision and robust state policy.

WHY THIS BOOK NOW?

The past decade has seen great advancements from China in the space
domain. If this trajectory continues unchallenged, the United States will
soon find itself at a serious economic and military disadvantage. Unfortunately, the United States has been pursuing space policy and objectives
without any long-term strategy. Various American policymakers have
advocated for space to different degrees, but none has done so in a truly
comprehensive, coherent, and vigorous manner. This project aims to provide a road map for preventing China from assuming the mantle of the
world's preeminent space power—and to position the United States as the
technical and moral leader of the world through superiority in space.

Building on the knowledge of space leaders in the U.S. military, government, academia, and the private sector, this book benefits from expert
insights across all segments of the space industry and provides recommendations for policymakers to use that span multiple presidential administrations. It is intended to serve as a guidebook for legislators and executive
branch officials who will confront the space issues outlined in these pages.
In an era of increasing space development, it is imperative to understand
not just military threats to the U.S. space architecture but also the challenges posed by adversarial space programs, opportunities for private sector and government cooperation, and the unbounded economic potential
space offers.

This book will help readers to understand the full landscape of space
policy, detail the threats to U.S. space assets and the economic advantages
of the burgeoning space economy, and offer solutions to some of today's
most pressing space-related policy challenges. More than a simple history
or summary of where we stand, the volume will look ahead to debates that
are only now cresting the wave of public policy discourse and promise to
be relevant over the next decade and beyond.

CHAPTER 1

Space Is an Untapped Resource

Peter A. Garretson, Richard M. Harrison,
and Anthony Imperato

In the decades following the Moon landing, NASA initially steered America's focus in the space domain with a primary objective of space stations and Earth observation and then later exploration. Today, space has become essential for modern society. Satellites enable near-instantaneous and ubiquitous communication, high-precision global navigation, rapid financial transactions, and improved weather forecasting, among many other innovations that society now relies on. The U.S. military, meanwhile, has reaped the benefits of space for secure global communications, intelligence collection, ground forces positioning, and weapons guidance—all because of robust satellite architecture. However, these developments in space barely scratch the surface of what is achievable. NASA spin-offs, or technology derived from space missions, have brought significant benefits to society over the years, ranging from firefighter suits to memory foam to water filtration to technology found in computed tomography and magnetic resonance imaging scanners.[1] Moreover, the advent of reusable rockets and advancements in artificial intelligence, 3-D printing, robotics, and other emerging technologies have made space more accessible and open for business.

What has been dubbed the "billionaire space race," accelerated specifically by the development of reusable rockets, has slashed the cost of carrying cargo into space by 85 percent over the past two decades.[2] Elon Musk, Jeff Bezos, and Richard Branson are visionaries who believe in humankind becoming a multiplanetary spacefaring species. They also understand the value of space and its ability to be commercialized. Now is the time to shift from a focus on space exploration to one of space commerce, space tourism, and space development. For long-duration operations in space, it will be imperative to devote resources to nuclear power and propulsion systems. Solar energy will be indispensable for further development in space and for providing power to Earth. And asteroid and Lunar mining, along with in-space manufacturing, will provide the resources needed to fully realize all that space has to offer humanity.

Before the Trump administration, the U.S. government had not championed the importance of space, and American space objectives previously vacillated between presidential administrations. Somewhat surprisingly, a transfer of power to the Biden administration has not steered the positive trend in U.S. space policy off course. Following a series of executive orders by the Trump administration, the Biden administration's National Space Council has continued on the path to space modernization with the release of a U.S. Space Priorities Framework, which mostly tracked with the previous administration's space guidance.[3] The U.S. government can no longer ignore either its reliance on space assets for military and civilian use or the potential advantages of a space economy, already worth $450 billion and doubling less than every 10 years, with a potential to grow to trillions of dollars over the next few decades.[4]

EXPANDING GLOBAL INTEREST IN THE SPACE ECONOMY

While the exact magnitude of the space economy is up for debate, there is now broad consensus that the space economy is poised to grow significantly. Major financial institutions have provided forecasts of what that economy might look like by 2040. Goldman Sachs has estimated the projected value of the space economy at $1.1 trillion; Morgan Stanley reached the same figure.[5] Similarly, Citigroup places the space economy value at over $1 trillion at that time.[6] The U.S. Chamber of Commerce, meanwhile, pegs the value even higher: $1.5 trillion.[7] The figure projected for 2050 is even more impressive—estimated by both Bank of America and Merrill Lynch at $2.7 trillion annually (for comparative measure, this is only a fraction of the $10 trillion value estimated by China at that time).[8] Reflecting this optimism, former U.S. secretary of commerce Wilbur Ross gave a speech at the 2020 World Economic Forum in Davos, in which he noted that "current industry projections place the 2040 global space economy at between $1 and $3 trillion. And I think we will certainly get to a trillion before 2030."[9] Ross specifically mentioned America's near-term priorities in this domain to include Lunar mining, asteroid mining, space tourism, and remaining the premier space power. Former vice president Mike Pence has similarly said, "[I]n this century, we're going back to the Moon with new ambitions, not just to travel there, not just to develop technologies there, but also to mine oxygen from Lunar rocks that will refuel our ships [and] to use nuclear power to extract water from the permanently shadowed craters of the South Pole."[10] What may sound like science fiction is quickly becoming a reality as technology continues to evolve. Vice President Kamala Harris is equally impressed with the potential of space, commenting that "space activity is education. Space activity is also economic growth. It is also innovation and inspiration. And it is about our security and our strength."[11]

Some space experts believe that these predictions are conservative in nature and underestimate the space economy's potential. SpaceFund cofounder and managing partner Meagan Murphy Crawford has argued that the space economy will soon become a multi-trillion-dollar sector and that the economic potential of space is limitless.[12] A key factor contributing to Crawford's prediction is the commercial space sector's efforts to expand access to space. For example, a number of private space companies recently have announced initiatives to develop commercial space stations. The Houston-based private space company Axiom was selected by NASA in January 2020 to develop the commercial successor to the International Space Station (ISS). In October 2021, Blue Origin and Sierra Space announced a partnership with Redwire Space, Genesis Engineering Solutions, Boeing, and Arizona State University to develop the Orbital Reef commercial space station. Additionally, Nanoracks also announced in October 2021 a partnership with Lockheed Martin and Voyager Space to develop the Starlab commercial space station. The development of these commercial space stations has the potential to allow for expanded space tourism and revolutionary research and development.

George Pullen, chief economist of Milky Way Economy, instructor in space economics at Columbia University, and a senior economist of the U.S. Commodity Futures Trading Commission, has developed projections that are similar to the estimates of the People's Republic of China (PRC) of $10 trillion by 2050 (see figure 1.1). The data suggest that by 2040 the space economy could be 4 percent of global gross domestic product (GDP), rising to 9 percent of global GDP by 2070.

The founder and CEO of Trans Astronautica (TransAstra) Corporation, Joel Sercel, has likewise argued that current space economy projections are massively understated and that in the coming decades the space economy will account for a large portion of the overall terrestrial economy.[13] Sercel argues that as the space industry develops reusable space vehicles, space launch and travel costs as well as orbital logistics costs will be comparable to or even less than aviation costs. Such a development would allow for Lunar and asteroid mining and for the in-space manufacturing of terrestrial products and space-based assets. As a result, space would become an even more integral part of the global economy.[14]

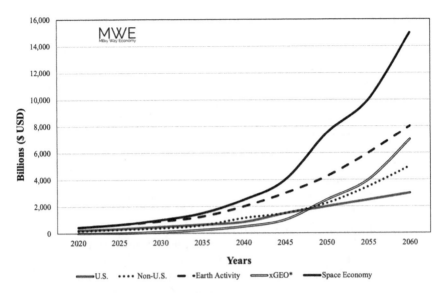

Figure 1.1 Space Economy, 2020–2060 Estimates
*xGEO = outer space beyond Earth orbit
Source: Milky Way Economy.

CITIZEN SPACEFLIGHT ("SPACE TOURISM") AND SPACE SERVICES

The success and technological prowess of commercial space companies—such as SpaceX, Blue Origin, and Virgin Galactic—has led to the emergence of space tourism. SpaceX made history in September 2021 by sending a crew of four private astronauts into orbit aboard the company's Crew Dragon capsule.[15] Blue Origin sent its founder, Jeff Bezos,[16] and celebrities such as NFL Hall of Famer Michael Strahan[17] and *Star Trek* star William Shatner[18] to suborbital space in 2021 as well. Additionally, Virgin Galactic launched its founder, Richard Branson, to suborbital space in July 2021.[19] Thanks to the efforts of these companies, their future initiatives, and the commercial space sector more broadly—together with the development of reusable rockets—the global space tourism industry is poised to grow considerably over the next decade.[20] While precise estimates vary, the value of the space tourism market in 2020 was assessed at $651 million, according to Global Industry Analysts Inc.[21] UBS estimates that the space tourism market is expected to reach $4 billion by 2030.[22]

The growth of space tourism will lead to truly revolutionary developments for humanity in space, such as the construction of private space stations ("space hotels"), which will potentially allow people to vacation in orbit. Companies such as Axiom Space and Orbital Assembly Corporation (OAC) already have plans in place to develop space hotels. Axiom Space is in the process of developing the commercial successor to the ISS, known as Axiom Station, which will detach from the ISS upon its decommissioning.[23] Axiom Station will serve as a de facto space hotel, as it will be available to private space travelers.[24] The company is aiming to allow private space travelers to visit and remain at Axiom Station modules connected to the ISS by the end of 2024.[25] OAC is working to develop its proposed Voyager Station, which will rotate in low Earth orbit.[26] Depending on how the Voyager Station is assembled, it will be able to accommodate 316–440 passengers and will include a range of amenities such as a gym and activity center, restaurant, bar, library, and movie theater.[27] OAC plans to begin construction on the Voyager in 2026 and aims to have the station up and running by 2027.[28] The rotation and simulation of artificial gravity is a major step, as astronauts have quietly complained about the hardship of zero gravity on the human body for long durations in space.[29] Additionally, the Gateway Foundation is planning to construct the Gateway Spaceport in low Earth orbit. The spaceport would act as a transport hub for people as they venture out to other locations in space.[30]

For these reasons, the number of people in space is likely to grow. In a model by George Pullen, the number of people in space could exceed 350 by 2030 and 15,000 by 2050 (see figure 1.2).

In addition to facilitating space tourism, the advent of reusable space rockets with large capacity, such as SpaceX's Starship, will allow for point-to-point travel. Point-to-point travel describes the launch of a rocket into suborbital space, where it transits for a short duration before reentering Earth's atmosphere at a different location.[31] Point-to-point travel has the potential to provide immense commercial and military benefits by drastically reducing flight time and transporting people and cargo around the world in short order. As an example, SpaceX has said that its Starship would reduce the current 15-hour flight duration from New York to Shanghai to around 40 minutes.[32] The U.S. military has demonstrated interest in developing point-to-point transport capabilities, as it could quickly resupply troops in times of crisis during conflicts across the globe. In January 2022, the U.S. Air Force awarded a $102 million contract to SpaceX to develop and demonstrate point-to-point capabilities for military cargo and humanitarian aid.[33]

IN-SPACE SERVICE, ASSEMBLY, AND MANUFACTURING

In-space servicing, assembly, and manufacturing (ISAM) (previously known as on-orbit servicing, assembly, and manufacturing, or OSAM) is

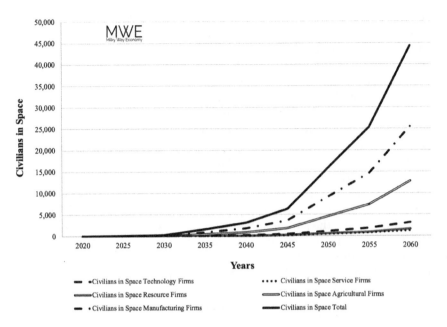

Figure 1.2 Population of Space, 2020–2060 Estimates
Source: Milky Way Economy.

an essential suite of tools for building in-space industry and an off-Earth supply chain. ISAM is among the fastest-maturing technologies and will benefit from the recently released White House national strategy advocating for building on these emerging capabilities.[34] Servicing and assembly will come first, with manufacturing not far behind. Coupled with space-based extractive industries (Lunar and asteroid mining), in-space power, and in-space logistics systems, these technologies open up a vast canvas of new industries and enable ISAM to a previously unthinkable scale.

Most satellites are treated as disposable—they are launched with a limited amount of onboard propellant, and for the most part, the satellites are neither refueled nor upgraded. A satellite's life expectancy is calculated based on maintaining its orbit and having enough fuel for a safe deorbit upon completion of the satellite's mission. However, satellites are now being developed with the capability to provide services to other satellites, like adjusting their orbits or delivering additional fuel, thereby extending the satellites' lives. Both in-space refueling and life-extension maneuvering have been demonstrated. For example, in February 2020, U.S. defense contractor Northrop Grumman's Mission Extension Vehicle-1 (MEV-1) completed the first successful life extension by docking with Intelsat 901 (IS-901) and moving it into a desirable orbit to extend the communication satellite's mission.[35] On-orbit robotic servicing translates to longer satellite lifetimes, upgrades, and greater maneuverability.

With the exception of manned space stations, like the ISS and China's Tiangong space station, most satellites and spacecraft are launched fully assembled. The size of satellites is limited by the size of the payload fairing—most satellites and spacecraft launch with their solar panels or other systems folded so they can fit in the small shells that protect the satellite during launch. Moreover, a major design constraint for satellites (and their complex unfolding mechanisms) is to be able to survive the forces of acceleration and vibration during launch. The ability to robotically assemble satellites in space unlocks entirely new designs, such as significantly more powerful and precise communication satellites, considerably larger in-space telescopes, large private space stations, and much more.

Today, each component of every space asset is manufactured on Earth using Earth-sourced materials. In-space manufacture will make possible the building of structural materials like trusses; the production of functional materials such as mirrors, coatings, photovoltaics, and integrated circuits to build satellites and facilities; and the manufacture of products that require microgravity or a vacuum for use on Earth (including ultrapure crystals, metals, glasses, fiber optics, and 3-D printed organs)—this market of microgravity research and development is estimated to be worth $14 billion in 2040.[36]

The future of ISAM is now. Two funded NASA missions (OSAM-1 and OSAM-2) will develop robotic abilities to build trusses and assemble spacecraft by about 2024.[37] NASA is already flying a 3-D printer on the ISS, helping companies to make high-end fiber optics and print organs.[38] Additionally, NASA has an entire program dedicated to in-space manufacturing. On Mars, the MOXIE experiment demonstrated in situ production of oxygen.[39] Several companies have plans to mine the Moon and produce feedstocks for ISAM as well as to manufacture power, structures, and functional components on the Moon as early as 2025. Moreover, through its NOM4D program, the Defense Advanced Research Projects Agency (DARPA) is funding multiple performers to develop technology to manufacture large optical, radio-frequency, and photovoltaics platforms in space and to use Lunar feedstocks.[40] Ultimately, the ability to manufacture structures and components, assemble them into complete systems, and service them over time opens exciting possibilities of entirely new services like space solar power (SSP) for large-scale space settlements and large spacecraft that cycle between planets.

ISAM is currently the cutting edge of space technology, and U.S. government investments will help ensure the competitiveness of our industrial base. Success in ISAM opens up entirely new possibilities for civil space exploration, such as better architectures for human exploration and astronomy. The advantages of ISAM will enable Space Force assets to refuel and maneuver without regret and will enable entirely new satellite designs in a broader diversity of orbits, contributing to resilience. The ability to assemble larger apertures (antennas and optics) with ever greater levels of electric power enables entirely new military capabilities like power beaming and space-based radar. Combined with space resources, ISAM can enable the resilience of an Earth-independent supply chain, possibly enabling reconstitution from space, even if an adversary were to deny the use of launch sites on Earth.

LUNAR ECONOMIC DEVELOPMENT

The vast resources of Earth's Moon will be the natural starting place to begin ISAM, before moving on to harvest the trillions of asteroid feedstocks. The ability to make use of Lunar feedstocks to build new industries and enable space industrialization is one reason why NASA has been directed to return to the Moon and specifically to the South Pole of the Moon. The South Pole is important because it contains significant water (in the form of ice) in its craters and gets nearly constant sunlight (for electric power) at the crater rims. The Lunar regolith (dirt) is full of aluminum, oxygen, iron, silicon, magnesium, titanium, potassium, phosphorous, and other desirable materials.[41]

Artemis is the U.S. program to return to the Moon. NASA's goals are to return boots on the surface of the Moon with a woman and person of color (a historical first) by 2026[42] and then to establish a permanent presence on the South Pole. The formal program must get astronauts from Earth's surface to the Lunar surface and back. The program's components include a giant heavy-lift rocket called the Space Launch System (SLS); the Orion capsule for the astronauts; the Gateway space station; and the Human Landing System (HLS)—a variant of SpaceX's Starship designed to take astronauts from the Gateway to the Lunar surface and back. Artemis is supported by the Commercial Lunar Payload Services (CLPS) initiative, which is funding small commercial landers to survey and emplace advance payloads. Artemis will put in place the basic logistics and power infrastructure to allow industry to grow and will make available the feedstocks of metals and glasses needed for in-space manufacturing and assembly as well as the water needed as a space commodity and oxygen needed for in-space refueling.

SPACE MINING

Space mining is the ability to access raw materials from asteroids, moons (starting with our Moon), planets, and comets. Developing proficiency in this extractive industries space sector could give the United States a significant economic advantage for a number of reasons.

First, Earth's gravity well is very deep, and it is extremely costly to bring anything up from Earth. In contrast, it takes about 22 times less energy to escape the Moon's gravity and even less to escape an asteroid's gravity.[43] That means there is a significant force multiplier when you can procure resources in situ. The term *in-situ resource utilization* refers to the use of resources that already reside in space, or "living off the land." Today, we continue to build giant rockets to bring all the fuel and materials necessary for a mission in space, but the more efficient option would be to access the fuel and building materials from the Moon or an asteroid. Not to mention that mining dead rocks avoids the ecological damage of mining on a living planet like Earth. Citigroup estimates the Moon will be mined first and that Moon mining will be worth $12 billion in annual sales by 2040.[44]

Second, the resources of space are already ideally located for ambitious projects like off-world bases, hotels, factories, and solar power satellites.

Third, the quantity of useful materials and sheer wealth in space is unfathomable, valued at about $700 quintillion. According to a NASA report, "It has been estimated that the mineral wealth resident in the belt of asteroids between the orbits of Mars and Jupiter would be equivalent to about $100 billion for every person on Earth today."[45] A 2020 *Wall Street Journal* article provided a window into the dividends that space mining just one asteroid could yield:

There's a lot of wealth in space. A 79-foot-wide asteroid could hold 33 tons of extractable material, including $50 million worth of platinum. The 2-mile-wide asteroid 1986 DA could be worth $7 trillion. But that will require massive investment in new technology, and investors need assurance that they won't pour billions into capturing an asteroid or mining the Moon only to be told the resulting product isn't theirs.[46]

Fourth, space mining would enable extraction of extremely valuable materials that are rare on Earth. Rare earth elements are vital for use in communications systems (such as mobile phones), electric vehicles, and advanced military systems including weapons. The United States does not currently have mining and processing infrastructure; rather, we rely on China, which has a stranglehold on the supply chain.[47] There are, of course, competing terrestrial solutions to reduce dependencies on rare earths, including recycling and technological advancements from NASA in advanced ceramics and additive manufacturing, but the concentrations and scale of the asteroid resources must be considered.

Once an asteroid—or an area of the Moon, for example—is identified for mining, there are several applications for the use of the mined material. The first is to support a space outpost with oxygen, water, habitat-building materials, road-building materials, and rocket fuel. Currently, the best cost to carry supplies to low Earth orbit has dropped from the Shuttle-era price of $54,500 per kilogram (kg) to $2,720/kg for a SpaceX-made Falcon 9

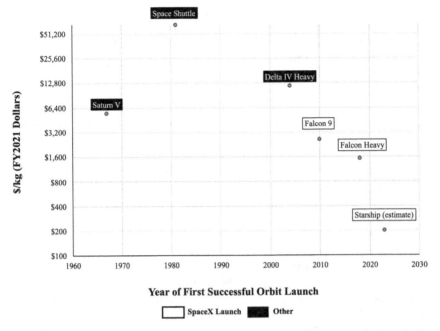

Figure 1.3   Medium and Heavy Space Launch Costs per Kilogram to Low Earth Orbit

Sources: Inspiration for the chart was from a similar graphic by Futureblind, https://futureblind.com/category/business/; Data used to develop the chart can be found on the Aerospace Security Project website at the Center for Strategic and International Studies, https://aerospace.csis.org/data/space-launch-to-low -earth-orbit-how-much-does-it-cost/.

rocket (see figure 1.3).[48] The cost to bring supplies to the Moon is \$15,000/ kg.[49] Therefore, any resource procured locally saves about \$15,000/kg in transportation costs.

Space-mined materials can also serve as a force multiplier for both exploration and defense. Sourcing propellants from the Moon or aster-oids enables the U.S. civil space program to travel much farther and with significantly more gear. Similarly, it enables U.S. military spacecraft to constantly maneuver and dodge, making a surprise attack much harder against them, while giving them the element of surprise (traditional crafts are limited by onboard propellant—space refueling is a very new phe-nomenon). As presently being explored in the DARPA NOM4D program, Lunar feedstocks may one day be used to construct large radio and opti-cal telescopes and photovoltaics to power them, enabling a larger, more robust defense satellite architecture above the reach of many threats.[50]

Furthermore, space-mined materials can be utilized for in-space man-ufacturing of rare, valuable items. These might include feedstocks for specialty fiber optic cables, specialty glass such as sapphire glass, and metal crystals (such as for jet engine blades). In extracting oxygen from the metal oxides in the regolith, metals are readily available as a by-product; it will be possible to use the oxygen, metals, and slag to build out the habitat and Lunar industrial facility to enable greater scale production. It is then possible to mine structural materials, like aluminum and iron to build spacecraft and structures, and high-tech parts for use on Earth, such as jet engine blades or superhard aircraft canopies. In the process, vacuum and low-gravity processes will enable extreme purity for applications like the manufacture of lithium batteries. Along the way, while materials are being separated, it will likely be possible to find high concentrations of platinum group metals (PGMs), which are critical minerals for the green economy and integrated circuits and are strategic minerals for national defense. PGMs may be valuable enough to ship to Earth. According to a report from Citigroup, "[F]or transporting to Earth, heat shielded capsules filled with gas bubbles could also be used and simply dropped into the oceans for ships to tow them away."[51]

And as mentioned earlier, space-mined materials are desirable for the prized rare materials themselves. Aside from PGMs essential for high-tech devices, fuel cells, and the hydrogen economy, one of the most valued commodities may be helium-3 (He-3)—a rare isotope in the solar wind that embeds itself in the Lunar dust deflected by Earth's magnetic field. When humanity finally succeeds in sustained fusion, He-3 will be a highly desirable fuel both for fusion power on Earth and for fusion power and propulsion in space because it produces very few neutrons, which are harmful to humans and damage reactors. Approximately 25 tons of He-3 (which would fit in the space shuttle's cargo bay) has enough energy to power the entire United States for a year and, therefore, is worth about $3 billion per ton.[52] In addition to America, both China and India, with vast populations to sustain, have an interest in this potential clean energy fuel.

Finally, the largest market and use for space mining will likely be entirely new industries built in space.[53] These may include construction of very large satellites, large space stations, factories, server farms, and power stations. For example, global energy needs are forecast to grow from 31 terawatts (TW) to 55 TW by 2100.[54] A single solar power satellite (discussed below) might have a mass of 10,000 metric tons, and more than 90 percent of the materials required for its construction (silicon, aluminum, iron) are available on the Moon. With today's launch economics, less than 4 percent of a rocket from Earth can be cargo.[55] However, with significantly less gravity to overcome, a launch from the Moon can carry 50 percent cargo—with zero impact to the Moon's biosphere, though it will be important to ensure the foreign environment is not damaged by mining and launches as well.[56] Eventually, because the Moon has no atmosphere, an electric catapult may be more economical than a rocket.

Regarding the technology to mine the Moon, the process has many similarities to mining on Earth—though it must be conducted in a vacuum and must cope with temperature extremes and the very sharp and fine Lunar dust (the Moon mining enterprise is estimated to reach $12 billion by 2040).[57] The technology to mine an asteroid with almost no gravity at all is much different. One technique being pioneered by TransAstra is to capture an asteroid and "optically mine" it—concentrating sunlight to cause it to sputter apart and release volatile gases.[58]

The concept of space mining may sound like science fiction, but early space mining experiments will be happening as part of the United States' Artemis program and the Chinese-Russian International Lunar Research Station. Some companies are postured to start a basic capability about 2025 and rapidly expand through a bootstrapping approach. With proper enabling policy, the United States should expect hundreds of metric tons of metals, oxidizers, glasses, water, and slag by about 2030. Missing at the moment are any production targets or public-private partnerships for a Lunar industrial facility at the South Pole. Meaningful contracts to purchase extracted commodities, or a robust space commodities exchange where commodities and futures can be traded, are also absent.

Asteroid mining has the potential to scale quickly. The PRC has plans to capture a small asteroid and return it to Earth in the mid- to late 2030s. The small-satellite technology to survey asteroids is quite mature and has been demonstrated by governments. As mentioned previously, the U.S. company TransAstra has developed an optical mining technology to pulverize an asteroid and extract its volatiles,[59] as well as an "omnivore engine" to use a variety of volatile materials as reaction mass,[60] and a "Sutter Telescope System" concept to find small asteroids.[61] The first asteroids could be mined for propellant before 2030. The logistics force multiplier effect of using asteroid resources and commercial practices cannot be understated—it makes Mars and other ambitious deep space exploration programs affordable by cutting the cost fourfold.[62] Once the logistics force multiplier effect is established (i.e., "gas stations" in space), it allows the structural materials of the asteroid to be accessed efficiently. The mass return on investment (how big a spacecraft you must launch compared to the amount of material you procure) is quite staggering (28–70:1), and the scale of capital required for the first missions is well within what mining and petroleum industry companies currently pay, less than $3 billion, to develop projects with comparable return on investment.[63] Advancements in space-based solar power and nuclear power can serve as enabling technologies for vital activities including space mining.

The speed at which the United States enables asteroid mining is less a question of technology than it is a question of the organization of capital and whether there are anchor customers. Encouragingly, the U.S. government is beginning to foster more private-sector interest in space mining. President Trump issued an executive order, titled "Encouraging International Support for the Recovery and Use of Space Resources," to help encourage the private sector to conduct operations such as Moon and asteroid mining.[64]

The value of space mining is evident. However, whereas there are millions of asteroids, there is only one Moon orbiting around Earth, and there are major advantages to accessing high-value locations, such as at the poles for space mining. There are definitely first-mover advantages to securing the resources on the asteroids and the Moon, so development is critical for commercial operations, military advantages, and America's future in space.[65]

At this time, the question is not whether the wealth is in space or even if the technology to access the riches will be available within the next 20 years. Rather, it is whether there is confidence among investors that the efforts are worth the risk. The question is germane because Chinese space companies, which are ultimately tied to the Chinese government, will not have any hesitation to invest in space.

HARNESSING SOLAR ENERGY

The most promising and economically impactful application of the confluence of reusable space access, ISAM, Lunar development, and asteroid mining is SSP. Citigroup estimates the SSP industry will amount to $23 billion annually by 2040.[66] It is perhaps the most encouraging technology for improving the quality of life on Earth, to steward the biosphere, and to access the vast abundance of the solar system. Of all available energy sources, no green energy solution holds a candle to the abundance of power and longevity of the Sun, which has and will continue to produce a steady 430 quintillion joules of energy per hour for billions of years to come (for reference, all of humanity uses that amount in a year).[67] However, collecting that energy on Earth is no simple task. Terrestrial-based solar power collection, as currently configured (with solar panels on buildings, the ground, and the like), is not an efficient method for utilizing this great resource.

As sunlight hits Earth, it is unevenly distributed among tropical, temperate, and polar regions and loses about a third of its power just traversing the atmosphere. Solar energy is intermittent and varies greatly depending on location and requires vast tracts of land that create local heat islands to capture the power. Moreover, grounded solar produces much less than its rated amount in winter or when impacted by poor weather conditions, including dust and rain, and there is little capturable solar power in the morning or evening and, of course, nothing at night. This high variability and low-duty cycle require all of the above: a huge amount of overbuild, duplicate fossil fuel or nuclear backup systems, and vast amounts of energy storage. Additionally, the locations with the best sunlight are often distant from the areas where the power is needed, requiring expensive power transmission infrastructure that is also susceptible to attack and disrepair.

SSP avoids all these problems by placing the solar collectors in space (and beyond the shadow of the Earth). The energy return on investment and energy payback time appear very attractive relative to terrestrial alternatives.[68] The energy captured from constant, powerful sunlight is then beamed wirelessly with radio waves to antenna receivers on the ground. Energy could be sent directly to the point of need or could shift between areas of need as the load requirement changes. SSP presents the possibility of a U.S.-built and -owned green energy system appropriate for urban and industrial life: available 24 hours a day at constant and high-energy levels. SSP's ability to supply baseload power makes possible the choice of how much to rely on coal and nuclear. Even more exciting is that SSP appears to be able to scale to global energy demand several times over, even if the world population grew to 10 billion and even if all 10 billion people consumed energy like people in the United States. Thus, SSP presents a true technical and scalable solution to the linked problems of atmospheric carbon, climate change, sustainable development, and energy security. As mentioned in a report by McKinsey & Company and the World Economic Forum, "Moving industries like power production into orbit could play a role in reducing global warming and ensuring that Earth can continue to sustain human life."[69]

Today, the average energy cost in the United States, including residential, commercial, industrial, and transportation sectors, is approximately $0.11 per kilowatt-hour (kWh). By comparison, recent studies by NASA regarding its concept SPS-ALPHA (solar power satellite by means of arbitrarily large phased array) have estimated that "a full-scale SPS-ALPHA, when incorporating selected advances in key component technologies should be capable of delivering power at a levelized cost of electricity of approximately $0.09 per kilowatt-hour."[70] According to the report, the SPS-ALPHA would collect energy in space and beam it down to receiver antennas on Earth via microwave energy. Unlike gas, oil, and coal plants, SSP could provide energy to commercial, industrial, and residential users during peak consumption times with minimal environmental impact. The NASA study was conducted close to a decade ago, so with technological improvement over the years, revised estimates project that the SPS-ALPHA Mark III will produce electricity at an even lower price of $.06/kWh.[71] The NASA study and the follow-on analysis reach conclusions similar to those of the Japan Aerospace Exploration Agency study, which projected SSP energy rates at $0.065/kWh.[72] The costs for SSP are in line with today's electricity prices, but the assembly and development of space-based infrastructure to support SSP will be a major undertaking.

A single solar power satellite, producing power equivalent to a major nuclear power plant, is a megaproject by current standards. The structure would be many times the size and weight of the ISS and would require many launches to complete and assemble in space. To scale to all global demand, the United States, likely with allies, would need to build thousands of these solar power satellites.[73] The first generation of solar power satellite components would be built on Earth and provide a large market for reusable launch vehicles like the SpaceX Starship, Blue Origin's New Glenn, or Relativity Space's Terran R.[74] As SSP proves its viability, it will create market pull to make use of raw materials from the Moon and asteroids, expanding or shifting important parts of industry and minimizing the creation of pollutants that impact our biosphere, endangered species, and quality of life.

The possibility of sourcing material for solar power satellites (recognized by the United States' strategic competitors) adds a strategic dimension to the Moon. It strongly suggests that an important focus of the Artemis base camp should be the development of a Lunar industrial facility to mature procedures to source, process, manufacture, and launch raw materials and components for solar power satellites.

However, development and deployment of SSP will not be without difficulty or drawbacks. The "soft costs" of energy development and project deployment are as yet unknown. Critics of SSP have highlighted concerns about the transmission of radio-frequency beams heating the atmosphere; however, numerous studies show that these effects are negligible in comparison to the atmospheric heating produced by greenhouse gas emissions of fossil fuels.[75] Additionally, contrary to common belief, reports from the U.S. Food and Drug Administration do not show a link between microwave radiation and cancer.[76] Along the same lines, the National Institutes of Health's National Toxicology Program has not included pulsed or constant radio frequency as an identified cause of cancer in their report on carcinogens for Congress.[77] Though SSP wireless power beaming was not integrated in the outdated reports, concerns remain. However, SSP radio-frequency effects can be diminished through the use of receiver buffer zones and radio-frequency absorption panels located on receivers and via legislation regarding times of use, power-level restrictions, and receiver locations.[78]

While the idea of solar power satellites was originated, patented (now expired), and investigated in the United States, decades of government neglect have put us behind. Now China is in the lead, Japan is in second place, and the United States is trailing behind. Recently, however, the Department of Defense has taken a leadership role in the area. From a military perspective, SSP would be game changing, as it would provide expeditionary military operations and forward operating bases with a unique power delivery option that does not rely on easily targetable supply lines.

The U.S. Naval Research Laboratory recently flew the first-ever "sandwich tile"—a laptop-sized subcomponent that converts sunlight into radio waves on the secretive X-37B space plane.[79] The Air Force Research Laboratory has just ground-tested a much more advanced lightweight version, which they plan to launch on a near-term demo, as part of a plan to set a record of producing a one square meter solar photovoltaic to Radio Frequency phased array.[80] They have a path (but not the funding) to get to a megawatt-class demonstrator. Other efforts at NASA in ISAM and solar sails are helping to advance enabling technologies.

China expects to fly a near-term power beaming demo in the next couple of years and fly a megawatt-class demo (which will become the largest object in space, dwarfing the ISS) around 2028 (previously scheduled for 2030).[81] Both China and Japan assume they will have grid-competitive commercial systems by 2050. However, U.S. companies think they could achieve this milestone sooner, with anchor contracts, access to capital, and a favorable and enabling regulatory environment. The technology is mature enough that the United States could easily fly a megawatt-class prototype and service the first niche applications for government and remote industrial locations within a decade—if there was a national program structured as a public-private partnership similar to the Commercial Orbital Transportation Services program that created Falcon 9 and Dragon and made SpaceX among the most successful international space companies.[82]

NUCLEAR POWER AND PROPULSION

Power and propulsion are fundamentally enabling to reach deeper into space, undertake ambitious missions, and build a space economy. While solar energy is crucial for successful space development, it is only one piece of the energy puzzle.

Any sort of industrial civilization in space will require significant amounts of power for industrial processes, habitats, and transportation. Solar power has many advantages when relatively close to the Sun (such as in Earth orbit) and when unobstructed by shadow. However, wherever sunlight is not constant (such as on the Moon and Mars) or is weak (such as on Mars and in the asteroid belt) or where multiple reliable power sources are essential to life (such as in any human habitat), there are many reasons to prefer nuclear power. Similarly, in situations that require moving significant mass at great speed (such as human transport to Mars, interception of a dangerous asteroid or comet, or rapid military response between distant bodies), space nuclear propulsion is analogous to a jet engine versus a propeller. As such, the nation that leads in space nuclear power and propulsion is likely to be the leader in space exploration, space development, and space settlement.

There are several flavors of space nuclear power and propulsion (SNPP). They can be classified depending on the type of nuclear reaction: either fission (breaking apart large atoms to create energy) or fusion (forcing small atoms together to create energy). Alternatively, they can be classified depending on the application: propulsion, in-space power, or surface power (on a moon or planetary surface).

Fission is the most mature form of SNPP, and it can be divided into two major categories of power sources: radioisotope thermoelectric generators (RTGs) and reactors. Fission fuels are radioactive, meaning they naturally decay at a certain rate, splitting into smaller particles. Fission fuels can be used by themselves like a battery—when the small particles get thrown off, they create heat, and that heat can be turned into electricity. For decades, the United States has safely flown these nuclear batteries, called RTGs, to power its deep space probes.[83] Usually, these can produce only tens of watts to a few hundred watts.

A second, much more powerful technique is to build a reactor, where the natural decay from the fuel source splits other atoms in a chain reaction that causes the fuel to be used up much faster, but generating much more power—on the order of kilowatts or megawatts. On Earth, we have working fission reactors that supply our power grid, power our submarines, and have even been flown in aircraft. In space, the United States has flown only 1 nuclear reactor, in 1965.[84] Conversely, Russia flew about 30 during the Cold War.[85] Today, both China and Russia are working on developing space reactors.[86]

It is natural for laypeople to be concerned about launching a nuclear reactor. They may, for example, have heard of protests over the launch of RTGs and may assume that a significantly more powerful reactor presents a higher risk. In fact, the risk is considerably lower. The United States has an extremely stringent procedure* and has safely launched many RTGs, but because a space reactor is not ignited until after launch, a never-started space nuclear reactor is about 10,000 times less radioactive even than an RTG![87]

Nuclear reactors can be used for several distinct applications in space. First, space reactors can be used to provide heat and electrical power on a moon or planetary surface. This is the objective of the NASA surface power initiative, which seeks to place a nuclear reactor on the South Pole of the Moon to power the Artemis base camp before 2030.[88] The heat can help keep components and people warm during the Lunar night, turn ice to water, or provide energy for other industrial processes. The

---

* The ability of Chevron Deference or Chevron Doctrine may be overly restrictive for commercial interests. There are commercial entities that want to advance RTG technology, and current regulations may constitute a barrier.

electricity can keep the lights on, power industrial equipment, and recycle air and water.

Second, space reactors can provide heat and electrical power to space vehicles and space stations. The heat can keep components warm far from the Sun and can power instruments and communications for high-band-width data even at great distances. For space stations, nuclear reactors can power all the same components as on the Moon, enabling closed-cycle life support and industrial processes. Los Alamos National Laboratory and NASA recently developed a new reactor called KRUSTY, which could eas-ily fly in four years, but no program exists to fly it.[89] China has displayed great interest in the KRUSTY design and appears to have chosen a variant for its own space program.[90]

Space nuclear reactors can also be used to power superefficient elec-tric thrusters. That application is called nuclear electric propulsion (NEP). NEP is highly desirable for ambitious deep space exploration such as of Jupiter's moons. NEP is being pursued by China and Russia but not pres-ently the United States.

Third, space reactors can directly heat rocket fuels. Traditional rockets mix fuels (such as liquid hydrogen) and oxidizers (such as liquid oxygen) together to create great heat and a strong jet. Alternatively, nuclear ther-mal propulsion (NTP), also called nuclear thermal rockets (NTRs), pass propellant directly over the hot reactor, resulting in huge efficiency gains, essentially doubling performance. NTRs are actually simpler than chemi-cal rockets in design. As noted earlier, any kind of nuclear reactor can be safely launched because it produces significant radioactive material only after it is turned on (after achieving orbit). The United States developed NTP to a high degree of maturity during the Cold War—almost to flight demo—but never flew it. Today, both NASA and DARPA have embryonic NTP programs. The DARPA Demonstration Rocket for Agile Cislunar Operations program—funding permitting—hopes to fly the first NTP ever in 2025.[91]

Both NEP and NTP offer the possibilities of much faster transits (halving the time for a near-term system and potentially much shorter times rela-tive to a chemical rocket engine) between Earth and Mars or Earth and the asteroid belt, making it much safer for humans—there are concerns about the radiation humans can endure while transiting space, but the nuclear propulsion greatly decreases transit times relative to chemical rockets.[92]

Fusion is another exciting technology for power and propulsion. Today, numerous governments are attempting to advance fusion science (many traditional adversaries are even working collaboratively), and several companies are racing to create fusion reactors. In fact, venture capital has recently poured about $3 billion into the effort.[93] This sprint to achieve fusion is creating breakthroughs in plasma physics, materials development, and high-temperature superconductors. Many observers believe that fusion for space propulsion is technically easier than fusion for power. At least one company, HelicitySpace, has put forward a pathway to scale from an electric thruster with a modest gain from a few fusion reactions to significant gain and, ultimately, to a self-sustaining, self-powering fusion engine providing performance something like the engines in *The Expanse* (a science fiction television series portraying a future in which humans inhabit the asteroid belt and Mars). Unfortunately, currently there are no U.S. government programs seeking to advance fusion power and propulsion for space applications.

SNPP greatly extends the reach, sustainability, and speed of ambitious space projects. The Trump administration recognized the importance of nuclear energy and its applications for space in its executive order titled "Promoting Small Modular Reactors for National Defense and Space Exploration," but this vision has yet to be widely adopted and implemented across the government.[94] If America aspires to secure the advantage in space, it must invest in SNPP.

INVIGORATING THE SPACE ECONOMY

Although technology has not kept pace with popular expectations to date, the timeline of space technology is inherently fluid.[95] American economic progress in the space domain will be driven by politics and policy and by the seriousness with which we seek to address great power competition in this domain. The Pentagon has assessed that China is already pursuing this objective of space economic dominance with seriousness and a national commitment. We now face a national choice: Do we want to cede the strategic initiative, or are we prepared to forge a unified national strategy in response?

The rapidity of adoption of space technologies including solar and nuclear power, space mining, and ISAM could be exceptionally fast. The key question is whether the U.S. government is postured to play a facilitating role. Specifically, (a) is the government facilitating policy and vision? (b) is it investing sufficiently in fundamental technologies on a timeline and scale that is competitive with rivals? (c) has it created a one-stop-shop regulatory procedure? (d) is it signaling its interest clearly with publicly communicated challenges, architectures, and requirements? (e) is it acting as a lighthouse or an anchor customer to purchase commercial products and services to kick-start the market? and (f) has it facilitated low-cost capital available to encourage first entrants?

Chapter Highlights

- **The space economy is primed for development.** Major financial institutions forecast that the space economy will be in the trillions of dollars annually by 2040. Development of reusable rockets has slashed the cost of carrying cargo into space by 85 percent over the past two decades. It will be imperative to devote resources to nuclear power and propulsion systems. Solar energy will be indispensable for further development in space as well as to provide power to Earth. Asteroid and Lunar mining, along with in-space manufacturing, will provide the resources needed to fully realize all that space has to offer humanity.

- **New space services.** Point-to-point travel drastically reduces flight time, enabling people and cargo to be transported around the world for commercial and military benefit (e.g., a traditional 15-hour flight from New York City to Shanghai will take only 40 minutes).

- **In-space servicing, assembly, and manufacturing (ISAM).** Servicing allows for in-space refueling and satellite life extension. In-space assembly and manufacturing allow for larger, more complicated and durable structures.

- **Lunar economic development.** Feedstocks on the Moon can enable space industrialization. The Moon's South Pole contains water and ice in its craters, and Lunar regolith (dirt) contains aluminum, oxygen, iron, and other materials.

- **Space mining.** The mineral wealth resident in the belt of asteroids between the orbits of Mars and Jupiter has a value equivalent to $700 quintillion. Space-mined materials can be utilized for in-space manufacturing of rare, valuable items. The largest market and use for space mining will be construction of very large satellites, large space stations, factories, server farms, and power stations.

- **Space solar power (SSP).** Solar collectors in space (and beyond the shadow of Earth) collect constant powerful sunlight and then beam it round-the-clock wirelessly with radio waves to antenna receivers on the ground. SSP appears to be able to scale to all global energy demand several times over and provide energy during peak consumption times with minimal environmental impact.

- **Nuclear power and propulsion.** Power and propulsion are fundamental to the ability to reach deeper into space and build a space economy. Space reactors can provide heat and electrical power to space vehicles and space stations and power highly efficient electric thrusters on spacecraft.

CHAPTER 2

Competing with the Chinese Space Vision

Larry M. Wortzel

Today, China is pursuing a dominant position in space. It is doing so—not simply for exploration but as part of an effort to increase what the Chinese Communist Party (CCP) perceives as comprehensive national power (CNP).[1] This represents, among other things, an effort to build economic power, military power, diplomatic power and status, and international prestige, as well as to penetrate foreign markets. As part of this focus, China's leaders plan to make their nation a leading space power by 2045. The CCP is preparing to contest the United States in space with a road map that envisions the development of techniques for asteroid mining, the creation of nuclear-powered shuttles for space exploration, and the industrialization of the Moon to fabricate satellites that can harness energy in space and allow the Moon to serve as a base for further deep space exploration.

For China, accomplishing the long-range task of building the space industry represents a comprehensive enterprise involving conglomerates of state-owned enterprises in China, private or semiprivate enterprises, the Chinese People's Liberation Army (PLA), and the orchestrated efforts of the People's Republic of China (PRC) State Council.[2] As Dean Cheng of the Heritage Foundation has explained: "The way that the People's Republic of China (PRC) is governed allows it to pursue a more holistic approach to policy. The CCP not only controls the government, but also has a presence

The author thanks American Foreign Policy Council research interns Kyra Gustavsen of The Ohio State University, John Glenn School of Public Affairs and Andrew Hartnett of the George Washington University, Elliott School of International Affairs, for their assistance with research associated with this chapter.

in every major organization, including economic, technical, and academic entities. Consequently, China is able to pursue not only a 'whole of government' approach to policy, but a 'whole of society' approach."[3]

The increased CNP from its activities in space also helps to increase China's attractiveness to other countries that CCP general secretary Xi Jinping seeks to court via his Belt and Road Initiative (also known as the One Belt One Road Initiative, or BRI for short).[4] Communications and precision positioning, navigation, and timing (PNT) satellites allow China to sell telecommunications hardware, software, and commercial products to foreign nations. These systems may one day create an alternative to the U.S. Global Positioning System (GPS) for banking and financial transactions. And, as one author from the Chinese Academy of Social Sciences notes, economic and trade relations with Russia and space cooperation also contribute to China-Europe-Asia projects along the BRI.[5]

It is evident from China's "grand plan" for space that the development and funding of space infrastructure in many of the regions where the BRI is active is central to PRC planning. The entire range of terrestrial and space-based infrastructure being pursued by Beijing, in turn, is intended to increase China's economic, diplomatic, political, and military power.

CHINA: A SPACE POWER ON THE RISE

China is no newcomer to space. Its space program started in the late 1950s, at about the same time as its atomic bomb program, and as a reaction to the Cold War and China's ideological split with Russia.[6] Notably, it began as part of a national defense program. However, it was not until nearly half a century later that space began to assume a central role in the PRC's strategic planning.

The country issued its first white paper on space in 2000, and the concept of developing "high-capacity and long-endurance satellites for both civilian and military missions" first appeared in the PRC State Council's 10th Five-Year Plan (2001–2005).[7] Once the focus of China's state planning turned to space, it became a strategic priority for the country's leaders. Today, according to a report by the Project 2049 think tank, China's actions in space are "managed by a diverse set of military and civilian organizations, [and] Chinese political authorities view space power as one element of a broader international competition in comprehensive strength and science and technology (S&T)."[8] The PRC provided a 2021 perspective on its space program in a January 2022 white paper.[9]

In addition to being an economic and technological competitor to the United States, China has developed sophisticated capabilities to use spacepower to make its military more effective, creating a latent threat to the United States. Indeed, there is growing evidence that space warfare is seen by China's PLA as an integrated part of battle planning in any future conflict with the United States.[10]

The Chinese space program is rising quickly and is now poised to overtake that of the United States. By all important metrics, China has already eclipsed Russia to become the number-two global power in space. It has greater ambition in the space domain than does Moscow, as well as a larger budget, better organization, more robust industrial base, bigger commercial market and sector, and more numerous launches and satellites. China's successes in this domain have been aided by espionage as well as by weak export controls in the United States. For quite some time, U.S. space experts and legislators failed to recognize that scientific cooperation with China aided China's military and helped to create both a threat and a competitor.[11] While this reality is recognized more widely in the United States today, gaining a better understanding of the scope of China's space vision is essential in order for the United States to compete effectively in the space domain.

CHINESE MOTIVATIONS FOR SPACE

China's plans for space are wide ranging and include support for the BRI, economic as well as foreign policy development, mining for critical resources, exploration of the Moon and Mars, and establishment of its own space stations and bases. The PRC's "White Paper on China's Space Activities in 2016" painted its strategic objectives in terms of a five-year plan: "To explore the vast cosmos, develop the space industry and build China into a space power is a dream we pursue unremittingly. In the next five years and beyond China will uphold the concepts of innovative, balanced, green, open and shared development, and promote the comprehensive development of space science, space technology and space applications, so as to contribute more to both serving national development and improving the well-being of mankind."[12]

China's accomplishments—specifically, economic revenue, foreign policy advances, international prestige, and influence building—appear to be major motivators behind China's space program. As an example, the Chinese state-run *Xinhua* news agency documented some of the achievements in space cooperation with other countries. These include the following:

- Completing 51 space launches by the end of 2020.
- Placing into orbit 12 remote-sensing satellites developed by the Argentinian company Satellogic, with a projected future launch of a total of 90 satellites into orbit.
- Launching two satellites developed by Ethiopia.
- Completing agreements with 19 African countries to accept PRC television satellite coverage for a total of 8,162 villages.[13]

The subsequent white paper, "China's Space Program: A 2021 Perspective," sets ambitious goals for the next five years in applying newly engineered technology and tests of new space materials. According to the white paper, the areas China will develop in this period are "[s]mart self-management of spacecraft; a [s]pace mission extension vehicle; [i]nnovative space propulsion; [i]n-orbit service and maintenance of spacecraft; and [s]pace debris cleaning." To that end, China has announced plans to have the capability for heavy lift (more than 100 metric tons) and to convert its launch fleet entirely to reusable rockets. It also recently put forward concepts for a reusable two stage–to–orbit spacecraft that looks nearly identical to SpaceX's Starship and very recently stated it would like to use reusable rockets for its astronaut launches.[14]

These advances will accrue to the economic benefit of the PRC. According to the Institute for Defense Analyses, a U.S. defense policy think tank, China prioritizes "the value and importance of China's space industry as a high-technology industry and a means to promote indigenous innovation."[15] Of course, that does not tell the entire story. China has an active program to explore the Moon and beyond to Mars, to develop bases on the Moon, to explore the mining of the Moon and other space bodies, to operate the space station it launched into orbit, and—through a strategy of military-civil fusion (军民融合)—to "develop the People's Liberation Army (PLA) into a world class military by 2049."[16]

Why is China investing such effort in becoming the premier global space power? One strong motivation is a deep mistrust of the United States and its motivations in space. The authors of a recent book published by Current Affairs Press, a CCP-controlled publishing house, make clear that partnerships between U.S. industry and American national security space groups are worrisome.[17] The attitude in the PLA is perhaps best summed up by four faculty members of the National University of Defense Technology: they see the United States as seeking to preserve stability in space to protect its own interests while building a national space system that integrates military and civilian assets to support U.S. and allied joint space control.[18]

To a certain extent, this represents mirror imaging. As far back as 1956, CCP chairman Mao Zedong pointed out to the Politburo that there is an interrelationship between national economic construction and national defense building.[19] Thus, for the PRC, this type of collaboration among the industrial base, private enterprise, and national security space groups in the government is similar to the PRC's own military-civil fusion policy. Further, space developments are seen by some national security writers in China as systemic efforts in the United States to achieve dominance in all elements of the space domain and its ground-based infrastructure.[20]

As outlined in the next section, the PRC has a multipronged strategy that explores the steps to various facets of space. China's space agencies seek to create a rival to GPS and promote it for communications and PNT. As a step toward working with Russia to create a Lunar research station, China is working to establish and operate its own space station. With a view toward space mining for rare earth elements and other valuable minerals, China will explore Mars and other space bodies, seeking exploitable resources. And a major project has been designed to explore the possibility of transmitting solar power back to Earth and developing nuclear-powered space vehicles.

PRC Satellites Increasing in Number and Capability

A major indicator of the Chinese response to U.S. space dominance is the rapid growth in the number of satellites fielded by the PRC and the wide range of functions those satellites possess. As of January 2022, there are 4,852 known active satellites in orbit. Of these, 10.3 percent (499 satellites) are owned or operated by Chinese entities. This is more than twice the number of Russian satellites (169) currently in operation. The United States maintains an impressive 2,926 satellites, which is more than half of all known satellites in orbit.[21] Most are in low Earth orbit (LEO), including the PRC space station, but there are also systems in medium Earth orbit (MEO), in highly elliptical orbit (HEO), and in geosynchronous orbit (GEO) around Earth (see figure 2.1). Similar to U.S. private-sector company plans for mega constellations, China has announced plans to eventually put a constellation of almost 13,000 satellites in LEO, with the objective of improving broadband communications, including 5G communications.[22]

In addition, what may be a PRC ballistic missile launch system—similar to the U.S. Space-Based Infrared System (SBIRS),[23] which is designed to detect missile launches—is now in place. The system may be a precursor to a yet-to-be announced change in China's nuclear posture, moving the PRC from its long-declared "no-first-use" policy on nuclear weapons to a "launch-on-warning" policy. Such a change without any form of tacit or written agreement between China and the United States would raise the likelihood of vertical escalation in any confrontation involving long-range ballistic missiles between the two countries.[24]

Promulgating a Rival to GPS

China's BeiDou Navigation Satellite System is a major competitor to the U.S. GPS. Over 70 million smartphones in China are connected to BeiDou, and it will eventually be the PNT service of choice for some 120 countries associated with China's BRI. Seventy BRI member countries are already dependent on the system.[25] As of January 2022, there were reportedly 52 BeiDou satellites in China's constellation, surpassing the size of the U.S. GPS constellation.[26]

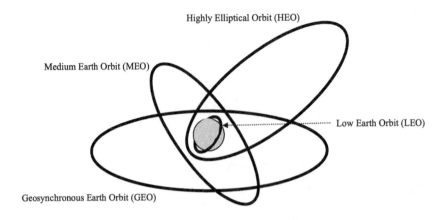

Figure 2.1 Orbits

Orbit	Altitude*	Uses
Low Earth Orbit	Up to 2,000 kilometers	- Communications - ISR - Human Spaceflight**
Medium Earth Orbit	Approx. 2,000 to 20,000 kilometers	- Communications - Positioning, Navigation, and Timing
Highly Elliptical Orbit	LEO altitudes at perigee (nearest to Earth) Approx. 40,000 kilometers at apogee (farthest from Earth)	- Communications - ISR - Missile Warning
Geosynchronous Earth Orbit	Approx. 36,000 kilometers	- Communications - ISR - Missile Warning

*The advantages of higher orbits for communications and ISR are near-persistent coverage of most of the Earth in view of the satellite, with the exception of Earth's polar regions where it is limited. LEO satellites cover all parts of the world, including the poles, but for shorter periods based on the speed of the satellite.
** With the exception of nine U.S. Apollo missions to the Moon, all human spaceflight has been completed in LEO.

Source: "2022 Challenges to Security in Space," U.S. Defense Intelligence Agency, April 2022, https://www.dia.mil/Portals/110/Documents/News/Military_Power_Publications/Challenges_Security_Space_2022.pdf.

BeiDou is a dual-use system. It serves as the major civilian PNT system for China, and simultaneously, all of the country's military services rely on it. One advantage of BeiDou is that it can transmit or receive up to 1,200 Chinese characters in a message, making it attractive to users.[27] Still, like any transmitting and receiving system, it is possible to track and locate a phone employing it, posing potential risks to users. China has plans to install a system of 1,000 ground stations to support BeiDou, and Chinese media claim that 98 percent of BeiDou components are made in China.[28]

The breadth of the BeiDou system, its utility, and the way that China is promoting it make it an economic challenge for the United States in a number of ways. Using BeiDou becomes an attractive means to communicate for cell phone owners or users in countries with inadequate cellular service, because it gives a satellite communications option, albeit a limited one. The U.S. system does not offer that alternative. Also, once a country adopts BeiDou, the PRC is able to set the standards for cellular communications and the form of system used in that country. As BeiDou increases in popularity and PRC standards become the norm, U.S. and European cellular systems will become less attractive.

Establishing a Foothold on the Moon and Beyond

The PRC has moved quickly through the stages of creating the infrastructure for launch to actually putting a space station into Cislunar space. The term *Cislunar* is Latin for the volume of space extending from Earth to the Moon and inclusive of the Moon's orbit and gravitational influence.[29]

The PRC has opted to construct its own space station, called Tiangong ("Heavenly Palace"). The effort will result in future commercial gains and prestige for China in addition to supporting scientific and technical research. The station is part of a broader effort to conduct operations on the Moon and explore deep space, including China's exploration and potential exploitation of resources on Mars.[30] Although there have been mishaps,[31] progress on Tiangong and its modules has proceeded methodically and according to plan for some time.[32] The planned space station system—which started in 1992 as an integral part of a crewed spaceflight, three-step process—is expected to be completed in 2022.[33] Tiangong will orbit about 400–450 kilometers (about 250 miles) above Earth in LEO.[34] The station is designed to last 10 years, with a crew of three astronauts (called taikonauts by the PRC). According to the *Wall Street Journal*, Tiangong will be about one-sixth the size of the "International Space Station, a 900,000-pound craft that over the past two decades has hosted over 200 astronauts from more than a dozen countries."[35] Like the International Space Station, Tiangong will have an integral robotic arm, and nearby plans call for a space telescope, like the U.S. Hubble.

Beyond Tiangong, the PRC has mapped out plans for exploration of both the Moon and Mars. In a 2016 BBC interview, Wu Weiren, the chief designer of the Chinese Lunar Exploration Program, outlined the country's future plans: "Our short-term goal is to orbit the Moon, land on the Moon, and take samples back from the Moon. Our long-term goal is to explore, land and settle. We want a manned Lunar landing to stay for longer periods and establish a research base."[36]

China's plans for the Moon also involve Russia: "China and Russia will use their accumulated experience in space science, research and development and use of space equipment and space technology to jointly develop a road map for the construction of an international Lunar scientific research station," according to reports of their space memorandum of understanding.[37] Russia and China will collaborate in the planning, design, development, and operation of the research station, which will be the largest space cooperation for China.

Like the United States, the European Union, and a number of other nations, China is seeking to establish a base on or near the far side of the Moon. In January 2019, China sent a Lunar probe to the far side of the Moon looking for places that would support the exploration of the inner solar system. However, to achieve communications and data exchange with Earth, the PRC had to launch a relay satellite with two other small satellites earlier, in May 2018. These early steps presage a Chinese push to establish an early foothold on the far side of the Moon. This is significant because, as other nations try to establish some type of base there as a launchpad for deeper space exploration, the number of suitable landing and basing sites may prove limited. Moreover, the area in which relay satellites can reside in ideal orbits above the Moon is limited—as other nations have already placed satellites at the Lagrange points that support satellite relay operations (for more details on Lagrange points, see chapter 3).[38] There is thus potential for a competition to access the best landing and basing sites. The nation that arrives first and establishes bases could enhance its chances for future deep space exploration and communications.

Beyond the Moon, China has set its sights on Mars. China's Mars exploration program has already successfully landed and communicated with a rover on the Red Planet.[39] According to the China National Space Administration, the Mars Zhurong (祝融) rover completed a planned 90-day mission in August 2021 and will continue to explore the planet in the vicinity of Mars's Utopia Planitia, the area where it originally set down. China reportedly aims to have "crewed missions to Mars by 2033 as part of a long-term plan to build a permanent base there."[40] This is part of a longer-term plan to construct a base on Mars and explore the planet.[41] But those plans experience periodic interruptions, because—depending on the position of the Sun, and whether it is between Mars and Earth—communications and transmissions between the two planets are cut off for periods of time.[42] Thus, over the next decade, without really knowing the scientific or potential economic value of being on Mars, China—like the United States, Russia, and the European Union—intends to spend a great deal of money, research, time, and effort in putting some permanent fixture there.

Notably, while some observers have proposed pathways for U.S.-China space cooperation on planetary exploration,[43] Chinese officials have articulated a substantially different view. According to China's chief Moon scientist, Ouyang Ziyuan, "[T]he Moon could serve as a new and tremendous supplier of energy and resources for human beings. . . . This is crucial to sustainable development of human beings on Earth. . . . Whoever first conquers the Moon will benefit first. . . . As for China, it needs to adopt a strategy based on its concrete economic power and technology level. . . . We are also looking further out into the Solar System—to Mars." Indeed, Chinese officials tend to see space in territorial terms, much like the PRC views the South and East China seas. "The universe is an ocean, the moon is the Diaoyu Islands, Mars is Huangyan Island," says Ye Peijan, the chief designer of the Chinese Lunar Exploration Program. "If we don't go there now even though we're capable of doing so, then we will be blamed by our descendants. If others go there, then they will take over, and you won't be able to go even if you want to. This is reason enough."[44]

Sino Space Mining

The PRC hopes that shortages of valuable, rare, or critical minerals can be addressed by exploration of the Moon and other space bodies. An area that appears to have potential is mining for rare earth elements on asteroids and the Moon. It bears noting that the United States has, in recent years, been caught off guard by China's dominance of the rare earth industry. Over time, the United States allowed itself to become dependent on what is now a potential adversary for some of the most crucial materials in high-tech production: rare earth elements.

Rare earths are a collection of 17 elements as essential to twenty-first century industry as oil was during the twentieth century. (The rare earth elements are scandium, yttrium, lanthanum, cerium, praseodymium, neodymium, promethium, samarium, europium, gadolinium, terbium, dysprosium, holmium, erbium, thulium, ytterbium, and lutetium.) They are important in producing a range of technological products, including cell phones, computer hard drives, and medical imaging equipment, as well as green technology like electric vehicle motors and wind turbines. In scientific circles in China, the management of mining, storage, and sale of rare earth materials is considered a major factor in national economic and foreign trade policy because of the country's dominance in industrial production.[45]

This significance has made space mining a topic increasingly under consideration in China. Over the past two decades, numerous papers[46] and government studies[47] have emphasized the importance and strategic value of space mining both in economic terms and as a way "to address China's oil and energy shortages."[48] Nor is the PRC limiting its vision of space mining to asteroids. Rather, it is also considering mining the Moon, Mercury, and Mars.[49]

Capitalizing on Space Solar Power (SSP)

China plans to put a mile-long stretch of linked solar panels into LEO by 2035.[50] Ultimately, the project is aiming to beam solar energy back to Earth to power a ground solar energy station that is to be fully operational by 2050.[51] Current plans call for the construction of a 1-megawatt solar energy station by 2030, which would be operational by 2035 and provide power at full capacity by 2050.[52] If the experimental project is successful, Chinese scientists hope that eventually a solar array in space would send as much electric power to Earth as a nuclear power station.[53]

Serious scientific articles about SSP and space-based systems are increasingly appearing in Chinese peer-reviewed journals. Several researchers are now looking into "harvesting solar energy by solar heating during the daytime and harnessing the coldness of the outer space through radiative cooling to produce electricity at night using a commercial thermoelectric module." Researchers have also developed a theoretical model for a device that would feature "24-hour electricity generation" and "off-grid and battery-free lighting and sensing."[54]

Nuclear-Powered Space Vehicles

The PRC is very interested in capitalizing on the benefits of nuclear energy for both power and propulsion in space. According to the *South China Morning Post*, "China is on course to develop nuclear-powered space shuttles by 2040, and will have the ability to mine resources from asteroids and build solar power plants in space soon after."[55] The PRC is planning a "nuclear fleet of carrier rockets and reusable hybrid-power carriers" to be ready for interplanetary flights and "commercial exploration and exploitation of natural resources by the mid-2040s."[56]

While satellites and spacecraft with nuclear power are nothing new, nuclear propulsion is a technology that has yet to be mastered. China has not yet launched a nuclear-powered satellite. But it clearly has ambitions to do so and expects that using nuclear-powered reactors will extend the life and perhaps the range of its satellites. There are also other possibilities, including weaponized versions of satellites. According to the *Space Review*, Russia is planning a new nuclear-powered satellite that could be used to power anti-satellite weapons or for electronic warfare from or in space.[57] In light of the competitive nature of China's approach to space, and the other military programs the PRC has in space, it is no surprise that China may follow Russia in attempting to create a weaponized, nuclear-powered space vehicle.

CHINA'S TERRESTRIAL INFRASTRUCTURE TO SUPPORT SPACE ACTIVITIES

Before introducing this topic, an important point must be made. The engineering, research, development, manufacturing, satellite or missile control, and launch facility infrastructures that support space activities are the same as that which support ballistic missile and warhead development.[58] That has always been a factor that influenced, for better or worse, how the United States approached any cooperation with China in space and also U.S. security and technology control programs.[59]

Another important consideration here is that the terrestrial infrastructure to support space was managed in China primarily by what was the General Armaments Department of the PLA. The tracking of missiles, warheads, and satellites—and monitoring of their respective telemetry and control systems in space, on land, and at sea—was managed by the General Armaments Department.[60] Today, with the elimination of the General Staff system in the PLA and the organization of the Central Military Commission Joint Staff Department, this infrastructure is probably managed by the PLA's Strategic Support Force.[61]

An updated list of the terrestrial space and missile support infrastructure inside China, as documented by the *China Brief*, includes:[62]

Launch Sites:
- Jiuquan Satellite Launch Center / Twentieth Testing and Training Base (63600部队) (中国酒泉卫星发射中心 / 第20试验训练基地)
- Taiyuan Satellite Launch Center / Twenty-Fifth Testing and Training Base (63710部队) (中国太原卫星发射中心 / 第25试验训练基地)
- Xichang Satellite Launch Center / Twenty-Seventh Testing and Training Base (63790部队) (中国西昌卫星发射中心 / 第27试验训练基地)
- Wenchang Aerospace Launch Site (文昌航天发射场)

Telemetry, Tracking, and Control Stations:
- Beijing Aerospace Flight Control Center (北京航天飞行控制中心)
- Xi'an Satellite Control Center / Twenty-Sixth Testing and Training Base (63750部队) (中国西安卫星测控中心 / 第26试验训练基地)
- Kashi station ([喀什- 航天测控站)
- China Satellite Maritime Tracking and Control Department
- Twenty-Third Testing and Training Base (中国卫星海上测控部 / 第23试验训练基地)

MILITARY APPLICATIONS OF CHINESE SPACE ACTIVITIES

Chinese views on the future of space as a battlefield are shaped by the way PRC military personnel view the United States. In a recently published book, two experts at the PLA National Defense University paint a dark view of U.S. activities in space and the establishment of the U.S. Space Force, while minimizing any reference to the way the PLA is using space.[63] The authors see the establishment of the Space Force and its activities as creating a new strategic situation and seeking to make the United States the dominant actor in space while preparing to utilize space to support military operations in all domains of war. Consequently, the authors postulate, China must respond and defend itself in that domain.[64]

To that end, China has established a wide-ranging system of ground-based interceptors, directed energy weapons, and space-based systems for anti-satellite operations.[65] Chinese publications have expressed concerns about the U.S. Air Force X-37B space plane, and China may have launched a similar space plane into orbit.[66]

All the space-related research and development in China is part of a state plan involving PRC universities and technology parks conducted under nationally mandated and managed programs to encourage basic research. But the research clearly has a military dimension as well.

In an illustration of the complexity and severity of the contemporary military threat from the PRC, China conducted two tests of nuclear-capable hypersonic glide vehicle (HGV) warheads in the summer of 2021.[67] A nuclear warhead launched in orbit that is deorbited toward a target—known as a fractional orbital bombardment—is not a new concept; the Soviet Union experimented with the idea in the 1960s.[68] But putting a warhead in orbit confers on an attacker a number of advantages, from an unlimited strike range to making early tracking of the warhead more difficult.

Whatever the particulars, the two tests highlight a new window of vulnerability for the United States.[69] Because China follows a countervalue nuclear doctrine, meaning a doctrine that seeks to inflict the maximum death and destruction on an enemy populace, even an inaccurate missile would cause catastrophic damage. A weapon like that, which could stay in orbit around the Earth for a considerable period of time, has the potential to evade all U.S. ballistic missile defenses, making it virtually certain the warhead will hit the United States.

Moreover, according to the *Financial Times*, the HGV was "being developed by the China Academy of Aerospace Aerodynamics," a research institute under China Aerospace Science and Technology Corporation, the same state-owned enterprise that designs missile systems for the PRC space program. And the test launch used a Long March rocket, the same rocket used to support China's space program.[70] These are telling indicators that the space program in the PRC is run by the country's military, as some congressional initiatives have contended.[71]

LEVERAGING THE BELT AND ROAD INITIATIVE AND SPACE

For some years, leaders in China have discussed the idea of a "space silk road" (天基丝路) and making that part of a "Belt and Road space information corridor."[72] The *South China Morning Post* notes that "providing financial support for African nations' space programmes is helping China to advance its soft power on the continent."[73] That is true of many of the destinations along the BRI, where China is including space-related programs in its infrastructure investments.

Africa provides a case in point. According to Julie Klinger of the University of Delaware, China is providing data sharing to African nations on "climate change mitigation, adaptation, and disaster response." These Earth-observation projects complement other programs on communications, satellite launches, and space technology development.[74] One of the "flagship examples" of what China can do for nations in space was a tracking, telemetry, and command station that it built in Swakopmund, Namibia, between October 2000 and July 2001.[75] With a space industry in Africa "estimated to be worth US$7 billion and projected to rise to US$10 billion" in another five years, China is ahead of the United States in developing inroads to take advantage of the financial possibilities and soft power that comes from its space program.[76]

Whereas Russian programs in Africa seek funding from the African nations to develop space programs, China usually covers the costs of the programs itself. For example, China paid $6 million of the $8 million cost to launch Ethiopia's first satellite.[77] And other programs continue in Sudan, Nigeria, and the Congo.

Another area of international cooperation for China is the countries of the Shanghai Cooperation Organization (SCO): the Russian Federation, Kazakhstan, Kyrgyzstan, Tajikistan, Uzbekistan, Pakistan, and India. China has leveraged its investments in these countries into the BRI, including space architecture, but its level of influence is constrained. The tensions and military and strategic competition between India and Pakistan may affect China's influence and infrastructure support to those nations. The fact that four of the SCO's member states were once an integral part of the Soviet Union, and a good deal of Russia's space and strategic missile production infrastructure is tied to Kazakhstan, also limits how far China can develop its own influence. This makes considerations of Russian policy a major factor in decisions made by the governments of Kazakhstan, Kyrgyzstan, Tajikistan, and Uzbekistan.[78]

Still, China has made progress with Pakistan, making it part of the "space silk road" and developing a long-term cooperation plan there.[79] In Kazakhstan, Kyrgyzstan, Tajikistan, and Uzbekistan, China has made some inroads in developing a space architecture related to the BRI, and Russia is not "the sole arbiter of policy" in the region. However, China's influence is still constrained by such factors as poor governance in the region and the business climate there.[80]

CHINA'S "SPACE ROAD MAP"

Today, China has largely accomplished the goals set out in its 2016 white paper on space, but it still has grand plans stretching out through 2045 (see table 2.1). Xi Jinping has declared that China's "Space Dream" is to overtake all nations and become the world's leading space power by that year. Looking ahead to 2045, the State Council has set out ambitious plans to focus on developing seven "frontier technologies."[81]

Specifically, in its five-year development plan, the 14th of its kind, Beijing said it would make "science and technology self-reliance and self-improvement a strategic pillar for national development" of key technologies on the frontier of the future, such as artificial intelligence, quantum information, integrated circuits/semiconductors, brain science, genomics and biotechnology, clinical medicine and health, and "deep exploration of space, the earth, the sea, and polar research."[82]

Table 2.1   China's Space Road Map (2020–2050)

2020	• China aimed to conduct its first Mars probe.[i]
2021	• China launches its first private reusable spacecraft.[ii] • Moon reconnaissance phase with Russia set to begin.[iii]
2022	• China will launch a probe to asteroids.[iv] • China will construct a space station.[v]
2024	• The Chang'e 6 will be used to bring back samples from the Lunar South Pole.[vi] • The Chang'e 7 will land a probe on the Lunar South Pole to survey it and search for ice.[vii]
2025	• China plans for space tourism using a reusable suborbital carrier.[viii] • First 100 kilowatt SSP demonstration at LEO.[ix] • Chinese and Russian space agencies are set to pick a site for the Moon base.[x] • Tianwen-2 will launch and seek out asteroid samples.[xi]
2026	• Construction for the International Lunar Research Station (ILRS) to begin.[xii]
2027	• The Chang'e 8 will launch to test technologies for incorporation in a Lunar base.[xiii]
2028	• Tianwen-3 will launch and return with a sample from Mars.[xiv] • China will launch a rocket capable of sending a crewed mission to the Moon.[xv] • Approximate year the Chinese economy is expected to pass the U.S. economy in real GDP.[xvi]
2029	• China to send a probe to Jupiter.[xvii]
2030	• China's Long March 9 super-heavy-lifter to be launched.[xviii] • China plans to demonstrate a megawatt-class space solar power in GEO.[xix]
2034	• China will capture an asteroid and return it to Earth.[xx]

Table 2.1   *(continued)*

2035	•	China will master key technologies such as space 3-D printing[xxi] and a 100-megawatt space solar power prototype.[xxii]
	•	The TianQin project aims to launch and deploy an equilateral triangular constellation with each side measuring 170,000 kilometers, forming a gravitational wave observatory. The project also aims to detect gravitational waves and draw a more comprehensive picture of the universe for humankind.[xxiii]
2036	•	Chinese settlers will land on the Moon and establish a Lunar research base. Habitats may be constructed using 3-D printing technology.[xxiv] China plans to send astronauts to the Moon by 2036 through the development of a superheavy carrier rocket, a crewed Lunar spacecraft, and a space suit suitable for a Lunar mission.
	•	The ILRS will become operational, providing a range of scientific facilities and equipment to study Lunar topography, geomorphology, chemistry, geology, and the internal structure of the Moon, as well as enabling space and Earth observations from the Moon's surface. It will also likely support human exploration in the future.[xxv]
2040	•	China will complete a nuclear-powered space fleet that will be ready for large-scale space mining and colonization.[xxvi]
	•	China's space-based solar power (SBSP) station will orbit above the Earth.[xxvii]
	•	The global space economy will reach a value of $1.1 trillion.[xxviii]
2045	•	According to Xi Jinping, China will be the most "advanced space nation."[xxix]
2049	•	The People's Republic of China will celebrate its 100th year and be the leading space technology power.[xxx]
	•	First commercial-level SBSP in operation in GEO.[xxxi]
2050	•	China will establish an Earth-Moon economic zone worth $10 trillion.[xxxii]

Note: GEO, geosynchronous orbit; LEO, low Earth orbit; SSP, space solar power.

[i] State Council of the People's Republic of China, "Full Text of White Paper on China's Space Activities in 2016," December 27, 2016, http://www.china.org.cn/government /whitepaper/node_7245058.htm.

[ii] Anqi Fan, "China's Reusable Suborbital Spacecraft Makes Successful Maiden Flight," *Global Times*, July 18, 2021, https://www.globaltimes.cn/page/202107 /1228956.shtml.

[iii] Tereza Pultarova, "Russia, China Reveal Moon Base Roadmap but No Plans for Astronaut Trips Yet," Space.com, June 17, 2021, https://www.space.com/china-russia -international-lunar-research-station.

[iv] Mike Wall, "China to Launch Ambitious Asteroid-Comet Mission in 2022," Space.com, April 18, 2019, https://www.space.com/china-asteroid-sample-return-comet-mission-2022 .html.

(continued)

Table 2.1 *(continued)*

v "China Launches Key Module of Space Station Planned for 2022, State Media Reports," CNBC, April 29, 2021, https://www.cnbc.com/2021/04/29/china-launches-key-module-of-space-station-planned-for-2022-.html.

vi "Future Chinese Lunar Missions," National Aeronautics and Space Administration Space Science Data Coordinated Archive (NSSDCA), https://nssdc.gsfc.nasa.gov/planetary/lunar/cnsa_moon_future.html.

vii Ibid.

viii Andrew Jones, "Chinese Company Aims for Suborbital Space Tourism With Familiar Rocket Design," Space.com, October 4, 2021, https://www.space.com/china-suborbital-space-tourism-cas-space-rockets.

ix David Brandt-Erichsen, "China Academy of Space Technology Continues R&D Into Commercialization of Solar Power Satellites," National Space Society, April 24, 2010, https://space.nss.org/china-academy-of-space-technology-continues-rd-into-commercialization-of-solar-power-satellites/.

x Pultarova, "Russia, China Reveal Moon Base Roadmap but No Plans for Astronaut Trips Yet."

xi Andrew Jones, "China to Launch Tianwen 2 Asteroid-Sampling Mission in 2025," Space.com, May 18, 2022, https://www.space.com/china-tianwen2-asteroid-sampling-mission-2025-launch.

xii Pultarova, "Russia, China Reveal Moon Base Roadmap but No Plans for Astronaut Trips Yet."

xiii "Future Chinese Lunar Missions."

xiv Jia Liu, "China Plans Two Mars Missions by 2028," Chines Academy of Sciences, September 20, 2018, https://english.cas.cn/newsroom/archive/china_archive/cn2018/201809/t20180920_197605.shtml; "Tianwen 2 Launch Forecasted for 2025! Spaceflight Window（134）(天问二号预计2025年发射！『航天视窗』（134）)," NetEase (网易), May 14, 2022, https://www.163.com/dy/article/H7ABM67K0531TTYW.html.

xv Ryan Woo, "China to Launch Rocket in 2028 Capable of Sending Crewed Probe to Moon," *Reuters*, September 29, 2021, https://www.reuters.com/world/china/china-launch-rocket-2028-capable-sending-crewed-probe-moon-2021-09-29/.

xvi "Chinese Economy to Overtake US 'by 2028' Due to Covid," *BBC News*, December 26, 2020, https://www.bbc.com/news/world-asia-china-55454146.

xvii Holly Chik, "China's Space Race Gathers Pace: Next Stop, Jupiter?" *South China Morning Post*, June 14, 2021, https://www.scmp.com/news/china/science/article/3137268/chinas-space-race-gathers-pace-next-stop-jupiter.

xviii Eric Berger, "China Officially Plans to Move Ahead With Super-Heavy Long March 9 Rocket," *Ars Technica*, February 24, 2021, https://arstechnica.com/science/2021/02/china-officially-plans-to-move-ahead-with-super-heavy-long-march-9-rocket/.

xix Chik, "China's Space Race Gathers Pace: Next Stop, Jupiter?"

xx "China Focus: Capture an Asteroid, Bring It Back to Earth?" *China Daily*, July 23, 2018, https://global.chinadaily.com.cn/a/201807/24/WS5b568b56a31031a351e8fbbd.html.

xxi Kubi Sertoglu, "China Celebrates Its First Set of 3D Printing Tests in Space," 3D Printing Industry, May 11, 2020, https://3dprintingindustry.com/news/china-celebrates-its-first-set-of-3d-printing-tests-in-space-171526/.

Table 2.1   *(continued)*

xxii "China to Build Space-Based Solar Power Station by 2035," *China Daily*, December 2, 2019, https://www.chinadaily.com.cn/a/201912/02/WS5de47aa8a310cf3e3557b515.html; Gao Ji, Hou Xinbin, and Wang Li, "Solar Power Satellites Research in China," *Online Journal of Space Communication* 9, no. 16 (Winter 2010), https://ohioopen.library.ohio.edu/cgi/viewcontent.cgi?article=1398&context=spacejournal.

xxiii Mario L. Major, "China Sets Its Eyes on Creating Laser That Cleans Up Space Junk," Interesting Engineering, January 2, 2018, https://interestingengineering.com/china-sets-its-eyes-on-creating-laser-that-cleans-up-space-junk.

xxiv Tia Vialva, "China National Space Administration to Establish 3D Printed Houses on the Moon," 3D Printing Industry, January 15, 2019, https://3dprintingindustry.com/news/china-national-space-administration-to-establish-3d-printed-houses-on-the-moon-147133/.

xxv Pultarova, "Russia, China Reveal Moon Base Roadmap but No Plans for Astronaut Trips Yet."

xxvi Jesse Johnson, "China Hopes to Build Nuclear-Powered Space Shuttle by 2040," *Japan Times*, November 18, 2017, https://www.japantimes.co.jp/news/2017/11/18/asia-pacific/science-health-asia-pacific/china-hopes-build-nuclear-powered-space-shuttle-2040/.

xxvii Zhao Lei, "Scientists Envision Solar Power Station in Space," *China Daily*, February 27, 2019, http://www.chinadaily.com.cn/a/201902/27/WS5c75c8b3a3106c65c34eb8e3.html.

xxviii "Space: Investing in the Final Frontier," Morgan Stanley, July 24, 2020. https://www.morganstanley.com/ideas/investing-in-space.

xxix Chi Ma, "China Aims to Be World-Leading Space Power by 2045," *China Daily*, November 17, 2017, https://www.chinadaily.com.cn/china/2017-11/17/content_34653486.htm.

xxx "Full Text of Xi Jinping's Report at 19th CPC National Congress," *China Daily*, October 4, 2017, http://www.chinadaily.com.cn/china/19thcpcnationalcongress/2017-11/04/content_34115212.htm.

xxxi Gao et al., "Solar Power Satellites Research in China"; Stephen Chen, "China Aims to Use Space-Based Solar Energy Station to Harvest Sun's Rays to Help Meet Power Needs," *South China Morning Post*, August 17, 2021, https://www.scmp.com/news/china/science/article/3145237/china-aims-use-space-based-solar-energy-station-harvest-suns.

xxxii Cao Siqi, "China Mulls $10 Trillion Earth-Moon Economic Zone," *Global Times*, November 1, 2019, https://www.globaltimes.cn/content/1168698.shtml.

The pronounced difference between the U.S. and Chinese space programs is that central planning has allowed China to construct a plan that will lead it to surpass America as the premier space power.

As the foregoing makes clear, it is impossible to separate China's civil space program from its military space program and its support for military operations by the PLA. Accordingly, Congress and the executive branch need to be wary of cooperation with China in this domain. Checks and balances must be established to ensure that commercial considerations do not significantly improve China's war-fighting capabilities. Moreover, it is necessary to acknowledge that China's progress in achieving its goals in space may proceed more smoothly than do U.S. efforts. The authoritarian nature of the Chinese state has created a streamlined—and ambitious—plan for dominance in the space domain. It is an approach, moreover, that is appealing to other authoritarian states and smaller nations that require financial or technical assistance to meet their objectives in space.

Chapter Highlights

- **Chinese space vision.** The CCP is preparing to contest the United States in space by developing techniques for asteroid mining, the creation of nuclear-powered shuttles for space exploration, and the industrialization of the Moon to fabricate satellites that can harness energy in space and serve as a base for further deep space exploration.

- **A space power on the rise.** China is the number-two global power in space. China's successes in this domain have been aided by espionage and weak export controls in the United States. American scientific cooperation with China aided the PRC military and helped to create both a threat and a competitor.

- **Chinese motivations for space.** China's plans for space are wide ranging and include supporting the Belt and Road Initiative, economic as well as foreign policy development, mining for critical resources, exploration of the Moon and Mars, and establishment of its own space stations and bases. The CCP also has a deep mistrust of the United States and its motivations in space.

- **Chinese space infrastructure.** The engineering, research, development, manufacturing, satellite or missile control, and launch facility infrastructures that support space activities are the same as that which support ballistic missile and warhead development.

- **PRC space military applications.** China has established a wide-ranging system of ground-based interceptors, directed energy weapons, and space-based systems for anti-satellite operations. Chinese publications have expressed concerns about the U.S. Air Force X-37B space plane, and China may have launched a similar space plane into orbit.

- **Leveraging the Belt and Road Initiative and space.** The "space silk road" (天基丝路) can provide financial support for African nations' space programs and advance China's soft power on the continent. China can build tracking, telemetry, and command stations for countries and cover the cost of development to gain soft power.

- **Chinese central planning.** The pronounced difference between the U.S. and Chinese space programs is that central planning has allowed China to construct a plan that will lead it to surpass America as the premier space power.

CHAPTER 3

Challenges to U.S. Space Security

Richard M. Harrison, Cody Retherford, and
Peter A. Garretson

Hostile nations are increasingly treating space as a war-fighting domain and pursuing strategic military activities that challenge American space security. The U.S. government and private sector can unlock the potential that space has to offer only if the operating environment is safe and secure. Space is still a foreign and hostile domain and must be secured from many threats, both natural and artificial. A brief and simple understanding of the threats is essential to develop strategies to secure the domain.

NATURAL THREATS (GEOMAGNETIC DISTURBANCES, ASTEROIDS, AND COMETS)

Militaries traditionally respond to adversary threats; however, they are also called upon to mitigate and respond to naturally occurring threats. Just as the United States must plan for natural threats such as hurricanes, tornadoes, earthquakes, and drought, it also must plan for space-derived natural threats. Doing so requires persistent environmental intelligence to provide advance warning and, in some unique cases, may even allow the U.S. military to avert such a disaster. Multiple naturally occurring threats that originate from the space domain require an enhanced U.S. situational awareness.

One class of threats is solar storms, or geomagnetic disturbances (GMD), including coronal mass ejections. These solar storms originate from the Sun and can send powerful bursts of charged particles at Earth, resulting in electromagnetic-pulse (EMP) weaponlike effects that can interfere with or even destroy communications and electrical infrastructure on the surface and unhardened space assets.[1] For example, 40 SpaceX Starlink internet communications satellites were lost in early 2022 as a result of increased atmospheric drag from a geomagnetic solar storm, which caused them to fall back to Earth just one day after launching.[2] No major harmful electromagnetic storm has occurred since we became an industrial, electrical society and civilization that could be seriously threatened by these events. However, past events such as the Carrington Event in 1859, and the more recent New York Railroad Storm in 1921, were so powerful that they set electrical fences and railway lines on fire. Civilization is much more vulnerable today.[3] Most protective measures are similar to hardening the electrical grid against the EMP threat, but there is a need for long-term environmental surveillance that provides advance warning of incoming solar threats. While such surveillance could reside in civilian agencies such as NASA, the National Oceanic and Atmospheric Administration, or the Department of Homeland Security's Federal Emergency Management Agency, militaries conceptualize domain awareness and surveillance differently from scientific organizations, which have a focus on discovery.

Earth impactors or Earth-crossing objects are another class of threats that must be addressed. These include asteroids and comets whose trajectories could strike Earth, creating damage ranging from broken windows from small (<50 meters in diameter) impactors that explode high in the atmosphere to city-killer size (~50 m) impactors to asteroids that could devastate whole continents (>150 m) or pose an existential risk or extinction-level event (>1 kilometer). While the population of potentially hazardous extinction-size asteroids is relatively well mapped (less than 10 percent remain undiscovered), few of the smaller asteroids have been found, and no last-minute-warning system exists.[4]

Even a relatively small meteor can cause significant damage and injury. For instance, the meteor that exploded over Chelyabinsk, Russia, on February 15, 2013, was only 17 meters wide but weighed approximately 10,000 tons and entered the atmosphere at 64,370 kilometers per hour (40,000 miles per hour). When it broke apart at an altitude of about 18 miles, the explosion was equivalent to 470 kilotons of TNT, significantly more powerful than a U.S. B61 nuclear warhead.[5] The explosion injured 1,200 people (none fatally) and damaged over 7,200 buildings. The Chelyabinsk meteor was referred to as "Earth's wake-up call."

Despite an order from Congress in 2008 to map these threats (Public Law No. 110–422), NASA has yet to meet the tasking more than 13 years later.[6] Moreover, unlike threats from asteroids whose orbits are close to Earth's orbit, long-period comets come from the farthest reaches of the solar system, are much harder to observe, are typically quite large, and provide at most a few months of warning. What is unique about impact threats is that they are the only class of major disasters that is preventable—with advance warning, they can be deflected or disrupted. However, despite congressional requests for the executive branch to recommend a lead for such mitigation, the mission remains unassigned. The failure to have a responsible agency, and the failure to invest in a deflection technology, constitutes a large risk to U.S. national security. The incorrect framing of the near-Earth asteroid problem as a mere science issue has marginalized the role that America's Space Force Guardians can and should play in homeland defense against the impact threat (see chapter 5).

In addition to the naturally occurring spaceborne threats (that are more akin to black swans), there are several threats already in the common orbit regimes. While space appears to be a great unbounded expanse, the area that Earth's satellites transit is actually fairly limited and is becoming increasingly cluttered. According to one Congressional Research Service report,

[O]ver 60 years of space activities—along with some explosive events in space including the 2007 Chinese anti-satellite (ASAT) test, the 2009 Iridium-Cosmos satellite collision, and India's ASAT test in 2019—have left large quantities of uncontrolled debris in these orbital "lanes." This includes tens of thousands of trackable items (softball size or bigger) and many millions (170 million according to NASA) of smaller objects, any of which may disable or destroy a satellite.[7]

In orbit, objects travel incredibly fast, so even a very small piece of debris can prove lethal to a satellite. Safely deorbiting old satellites and finding ways to mitigate orbital debris are critical to maintaining a safe space environment.[8]

ANTI-SATELLITE WEAPONS AND KINETIC AND NONKINETIC THREATS

Threats to U.S. space assets resulting from naturally occurring phenomena are predictive to varying degrees, and steps will need to be taken to address them. However, equally pressing are the threats from hostile adversaries, who are increasingly developing new and more sophisticated attack vectors that can harm U.S. space infrastructure. There are several types of counterspace threats, which can be classified into categories with varying destructive ability, including cyberspace and electronic warfare, directed energy weapons (DEWs), kinetic energy attacks, and orbital threats.[9] Threats to space infrastructure are not limited to attacks on satellites but can also affect ground stations on Earth and the communications links between them as well. Figure 3.1 illustrates the myriad counterspace threats.

In an ever-connected world, cybersecurity is necessary to protect network infrastructure, which includes satellites and ground stations. Hostile states can utilize nonkinetic cyber warfare operations to disrupt, degrade, or destroy connected space infrastructure.[10] For example, a cyberattack could involve hacking into an associated ground station infrastructure and manipulating a satellite's orbit.

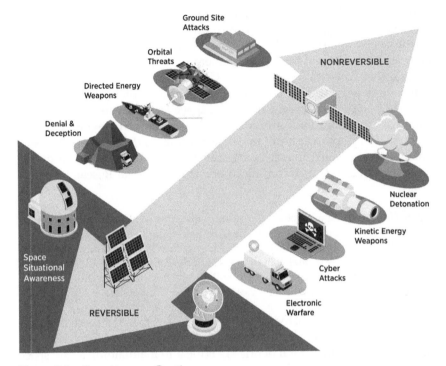

Figure 3.1   Counterspace Continuum
The counterspace continuum represents the range of threats to space-based services, arranged from reversible to nonreversible effects. Reversible effects, from denial and deception to electronic warfare, are nondestructive and temporary, and the system is able to resume normal operations after the incident. Directed energy weapons, cyberspace threats, and orbital threats can cause temporary or permanent effects. Permanent effects from kinetic energy attacks on space systems, physical attacks against space-related ground infrastructure, and nuclear detonation in space would result in degradation or physical destruction of space capability.

Source: "2022 Challenges to Security in Space," U.S. Defense Intelligence Agency, April 2022, https://www.dia.mil/Portals/110/Documents/News/Military_Power _Publications/Challenges_Security_Space_2022.pdf.

Electronic warfare (EW) is another nonkinetic attack vector that poses a threat to space systems by asserting control over the electromagnetic spectrum. Satellites are guided by, and in communication with, ground stations on Earth, and data are often transferred via high-frequency radio waves across hundreds or thousands of miles. Adversaries can attempt to disrupt signal transfer using techniques to block or jam the signal as it transits up from the ground station (uplinks) or down to the ground station (downlinks) or possibly between two satellites in orbit (crosslinks).

Additionally, hostile actors could send a false signal with erroneous data (known as spoofing) to deceive the intended receiver of the signal.[11] As an example, an adversary could spoof a signal to soldiers in a conflict zone by sending false navigation information.[12]

Similar to cyberspace operations, DEWs also have the ability to disrupt, damage, or destroy satellites by targeting sensor systems or a power source. Adversaries can use DEWs with a range of reversible to nonreversible effects including lasers, microwave, and radio-frequency weapons, in addition to the potential future use of particle beam weapons.[13] Most DEW beams travel at or near the speed of light, but the target must be in line of sight from the source of the weapon. These weapons can be deployed from Earth's surface or from aboard another satellite.

Kinetic energy weapons typically cause irreversible damage to satellite systems. Impacts from kinetic attacks can cause space debris, which puts additional satellites in similar orbits at risk. Anti-satellite (ASAT) weapons commonly refer to anti-ballistic missiles that have been altered to target and destroy satellites. These ASATs that rely on directly hitting a satellite with a kill vehicle, likened to hitting a bullet with a bullet, require extremely sensitive associated sensor and targeting systems. Additionally, rather than a hit-to-kill style, some ASATs rely on detonating a nuclear warhead in space to destroy the target satellite with the destructive blast or from the resulting electromagnetic pulse causing damage similar to that from a solar storm.[14]

While ASAT weapons are traditionally thought of as weapons launched from the ground or an aircraft, satellites can also face orbital threats from adversary satellites flying in similar or intersecting orbits. There are several payload types that could damage a satellite, including high-power microwave weapons, radio-frequency jammers, lasers, chemical sprayers (to foul payload optics and solar arrays), kinetic kill vehicles, and robotic mechanisms.[15] Co-orbital satellites are particularly dangerous because they can be dual use in nature; for example, a satellite used for on-orbit repair or to collect space debris could also be used nefariously to damage another satellite. These satellites may be placed in orbit through traditional launches or possibly distributed by a space plane or through other innovative methods.[16]

The aforementioned threats were described mostly with regard to on-orbit systems. However, threats to ground stations can be equally devastating during times of conflict. Ground stations may not always be located in a heavily guarded area, which could leave them vulnerable to adversary ballistic and even hypersonic weapons as they continue to gain prevalence.[17] Terrestrial-based infrastructure can be attacked directly and also potentially be disabled through cyber operations or through physical damage from attacks on the electrical grid or local power.

Hostile states are developing these offensive space capabilities, and their respective space programs pose a threat to U.S. national security. An April 2022 U.S. Defense Intelligence Agency report notes:

China and Russia are developing new space systems to improve their military effectiveness and reduce any reliance on U.S. space systems such as the Global Positioning System (GPS). Beijing and Moscow have also created separate space forces. As China's and Russia's space and counterspace capabilities increase, both nations are integrating space scenarios into their military exercises.[18]

While the major players are the largest concern, smaller actors have space programs that should not be ignored.

CHINESE SPACE CAPABILITIES AND DOCTRINE

China is executing a comprehensive national strategy aimed at achieving global space leadership by 2045 (see chapter 2). Unlike the United States, China's long-term goals are unaffected by elections; power shifts; or lack in funding, will, or interest. Beijing's space program is a point of national pride and is one of the means by which China will achieve military, intelligence, economic, and technological dominance over global competitors on Earth and control of resources in space. The Chinese space apparatus spans a variety of government agencies, state-run companies, and private-sector space companies, of which there are over 100 startups and established firms.[19] The China National Space Administration coordinates all government civilian space research and development; production; and space launch, exploration, and operations. The China Aerospace Science and Technology Corporation (CASC) and China Aerospace Science and Industry Corporation (CASIC) are state-run organizations that engage in the research, design, manufacture, testing, and launch of satellite launch vehicles (SLVs), satellites, and other space technology for civilian and military purposes.[20] The People's Liberation Army Strategic Support Force (PLASSF) executes both offensive and defense military space, cyber, and electronic warfare missions for China. These organizations coordinate with a variety of private-sector companies responsible for technological innovation and space asset development. Chinese companies such as LinkSpace, LandSpace, i-Space, Deep Blue Aerospace, and Space Trek seek to compete with American firms. While less advanced than U.S. government agencies and private-sector companies, public and private Chinese space ventures are rapidly accelerating to narrow the U.S. advantage.

Chinese space strategy focuses on accomplishing a set of long-term goals to compete with, and surpass, the United States by strategically emphasizing key areas that the United States neglects. The Chinese are attempting to gain advantage through investment in innovative emerging technologies to secure space-based energy (space solar power) and mineral resources (asteroid and planet mining), military and intelligence advantage, and dominance of the overall space economy; to conduct deep space exploration; and to aid in Lunar and Martian colonization efforts. They are investing heavily in new reusable rockets as well as satellite production facilities capable of producing hundreds of satellites per year.[21] The Chinese are working toward independent space-based telecommunications networks, including the recently completed BeiDou Navigation Satellite System, ending Chinese dependence on the U.S. GPS. Offensively, the Chinese are developing and deploying an array of weapons systems and satellites that will threaten American space assets and capabilities.[22]

Chinese Space Capabilities

The Chinese civilian space program is highly advanced and rapidly expanding to compete with that of the United States. Often, under the watchful eye of the government, Chinese private-sector space companies and startups are spreading rapidly and supporting expansion of the Chinese space program. Their efforts, combined with the two major state-run firms, CASC and CASIC, are leading to the development of innovative satellites and spacecraft, which are pushing space exploration forward.

The Chinese have put in orbit and are operating more than 320 satellites, providing GPS, communications, intelligence, and remote-sensing capabilities, and they are working to expand their Lunar and Martian exploration and colonization efforts. The space capabilities of the People's Republic of China (PRC) depend on a variety of SLVs, including the Long March LM-2 for medium-lift missions and the LM-5 for heavy-lift missions, and they are also developing the reusable LM-8 for medium-lift and the LM-9 for super-heavy-lift missions.[23] The four Chinese land-based launch facilities completed 42 launches in 2020, while the De Bo 3 sea-based launch platform completed 1 successful launch. During 2021, China set a record of 55 launches, over a dozen more than previous years.[24] Several of these launches carried mainly satellites and segments of the Chinese space station to support its construction.[25] The space station is expected to provide long-term space-based, human-executed research and development capabilities—unlike the previous Chinese space stations, which were short-term technology demonstrators. China further plans to build a permanent Lunar research station by 2036, allowing for the extraction of Lunar resources.[26] In addition to space research facilities, Beijing completed its fifth Lunar exploration mission, Chang'e 5, in December 2020 and successfully landed the Tianwen-1 rover on Mars in May 2021 to advance China's Martian exploration efforts in support of future human space stations.

The Chinese are also using space-based assets with dual-use or purely military purposes to advance their global agenda on Earth. China maintains a considerable array of space-capable weapons systems including ASAT and ballistic missiles; co-orbital satellites; and directed energy, cyber, and electronic warfare capabilities.[27] The Chinese most famously tested their SC-19 ASAT missile against a satellite in January 2007, but they have conducted multiple ballistic and likely ASAT missile tests in recent years and created ASAT missile units in the PLASSF. Furthermore, after assessing how Elon Musk's Starlink satellite system supported Ukraine during the conflict with Russia in 2022, the PRC is now considering developing a "hard kill and soft kill" solution to counter a decentralized megaconstellation.[28]

In a less overt manner, the Chinese have targeted government agencies and aerospace companies with cyber espionage in efforts to collect information on U.S. capabilities. PRC cyber and electronic warfare capabilities have the ability to disrupt and degrade U.S. space assets and are rapidly advancing critical pieces of Chinese national security strategy. China is seeking to compete in the space plane market for military purposes. For example, CASC is constructing a reusable space plane similar in concept and mission to the X-37B—the Chinese space plane completed a successful two-day orbital test flight in September 2020.[29] Reports have also noted that China "tested a nuclear-capable hypersonic missile in August that circled the globe before speeding towards its target, demonstrating an advanced space capability that caught US intelligence by surprise"—this weapon is known as a fractional orbital bombardment (FOB) device.[30] Furthermore, CASIC is actively developing a separate reusable space station, the Tiangong, designed to carry astronauts, supplies, and smaller spacecraft into orbit—the craft is expected to be completed by 2025.[31] CASC is also developing a nuclear-powered space shuttle to support long-range and endurance space exploration missions with plans to be operational by 2045.[32]

Chinese Space Doctrine

The inseparably intertwined civilian and military space and rocket programs of China work together toward the common purpose of achieving global space leadership and global hegemony. Both components of China's space program have seen tremendous growth and innovation over the past decade, indicating serious changes in Chinese strategic thinking. Chinese intelligence and forced technology transfer strategy has also focused heavily on out-innovating and outmaneuvering the United States to steal research and plans to reduce costs for Chinese development efforts and also to counter U.S. space and counterspace systems.[33] These efforts directly support the People's Liberation Army's (PLA) overall space strategy and China's national goals set for 2045.

In 2015, the Chinese military declared space a military domain and made sweeping updates to their doctrine to place greater emphasis on strategic, asymmetric, and nonkinetic effects, including reorganization of forces and the creation of new units. Their doctrine is increasingly focused on multidomain warfare, which is based heavily on maintaining friendly information networks and degrading enemy networks. This led to the development of the PLASSF in December 2015 and the PLA Rocket Force (PLARF), a successor to previous ballistic missile units, which entered service in January 2016. Both maintain missions with space and counterspace implications. PLASSF operates GPS, communications, and intelligence satellites supporting PLA maneuvering, targeting, and kinetic activity as well as internal information, cyber, and electronic warfare units conducting offensive and defensive operations to support overall strategic objectives. PLARF operates conventional and nuclear ballistic, cruise, and hypersonic missiles of all ranges with the potential to threaten terrestrial and space assets. PLARF likely also manages China's newly created ASAT missile units. These units demonstrate China's desire to dominate preconflict competition and to deter serious military conflict in their efforts to become a global hegemon. Their space-based assets play directly into the "Three Warfares" strategy (public opinion warfare, psychological warfare, and legal warfare)[34] for preconflict competition by providing both information collection and suppression capabilities to support information warfare. Overall, the Chinese seek to maintain space-based technological and operational advantage over American space assets and capabilities to ensure victory in preconflict competition and war.

RUSSIAN SPACE CAPABILITIES AND DOCTRINE

Despite engaging in the Cold War, Washington and Moscow have at times engaged positively in the area of space. At other times, space has fueled some of the greatest competitiveness and innovation the United States has experienced. Currently, Russia has an increasingly active offensive space weapons program—the U.S. intelligence community assessed that Russia tested an anti-satellite weapon as recently as July 2020 and then again in November 2021.[35] Russia's space program goals may be less comprehensive than China's, but Russia's program remains an important one to monitor.

Russia is midway through execution of a 10-year national space strategy. In March 2016, the Russian government approved the Federal Space Program (FKP-2025) with a total budget equivalent to roughly $20 billion. FKP-2025 lays out detailed, strategic goals for various projects including satellite and SLV research, development, and implementation; completion of the Russian sections of the International Space Station; human Lunar exploration missions; and movement of their primary human space launch site from Baikonur, Kazakhstan, to Vostochny, Russia.[36] The Russian space program is a combination of state-run and private-sector space companies and the Russian military. Roscosmos State Corporation for Space Activities, a combination of former private space corporations and government agencies, was formed to nationalize space operations and remains Russia's primary civilian space exploration entity. Roscosmos is responsible for civilian research and development, spaceflight operations, and space exploration missions. The Russian Space Forces, a subordinate branch of the Russian Aerospace Forces, is responsible for all military applications of space assets and technology, including operating satellites and anti-space weapons. The space program is supported by private and public corporations including RSC Energia, NPO Energomash, NPO Lavochkin, NPO Molniya, Khrunichev State Research and Production Space Center, and the Progress Rocket Space Centre. These corporations are responsible for supporting Roscosmos and the Russian military with SLV, satellite, rocket engine, and ballistic missile research and development. The Russian space program, while inferior in some respects to the U.S. program, is highly capable and is working to reduce the U.S. advantage in space and counterspace capabilities.

The Russian Federation's space program has always been a method through which to challenge American power and influence around the world. While the Soviet space program was dominant in the early space race, the United States has surpassed Russia's efforts in the majority of space exploration fields. The Russians are seeking to reclaim their former space leadership and are utilizing their program to achieve military, intelligence, economic, and technological advantage over global competitors. The definitive medium- and long-term goals contained within Russia's comprehensive national space strategy demonstrate the nation's resolve to return to leading the international community as the preeminent space power. Like China, the Russian government maintains the political will to execute these plans. Unlike China, domestic interest and funding in Russia depend on the success of an economy highly reliant on natural resource extraction and sales. Funding and support for the space program, like many things in Russia, hinges on the success or failure of oil and gas sales and current sanctions against the state and the state's enterprises. Despite Russia's economic woes, it is intent on achieving its strategic goals in space exploration, utilization, and weaponization to directly threaten the U.S. economic and military advantage.

Russian Space Capabilities

Russia maintains a wide array of civilian and military space capabilities supporting its global strategic objectives. Space launch is provided by variants of the Soyuz SLV for medium-lift missions and the Proton-M SLV for heavy-lift missions. The Russians are also developing the Amur SLV, which will be partially reusable for medium-lift missions and the Yenisei SLV for super-heavy-lift missions. Russia utilizes its SLVs for both domestic, civilian and military, launches and for economic gain, providing commercial launch to customers around the world (with five launch facilities).

Outside of Russian territory, the Baikonur Cosmodrome in Kazakhstan remains Russia's primary launch facility for a variety of missions—the facility conducted seven launches in 2020. The Russians also conducted two launches at the European Space Agency's Guiana Space Center in French Guiana. Within Russian territory, the space program utilizes the Plesetsk Cosmodrome, which conducted seven launches in 2020 primarily for polar missions, and the Vostochny Cosmodrome, which conducted one launch in 2020. The Russians also launch out of Kapustin Yar but did not conduct any launches in 2020 from this location.[37]

Russian Space Forces and Roscosmos operate over 160 satellites providing GPS, intelligence, surveillance, and reconnaissance (ISR), communications, and co-orbital repair.[38] Of the satellites, 87 are used for communications, 34 are used for Earth and space observation, 29 are used for navigation and global positioning, and 18 are used for technology development. In addition to satellite operations, the civilian space program is working to expand Lunar exploration. The Russians intend to launch their first mission, Luna 25, in 2022 with a series of uncrewed missions over the next decade, leading to a human-led mission to the Moon.[39] Moscow is working to close the gap between U.S. and Russian reusable space plane technology. Russian shuttle designers also announced work on a space plane in March 2021.[40] While technologically inferior in key areas, including reusable rockets and space planes, Russia has long-term strategies that seek to rectify those issues and eventually surpass American technological advancements.

Russia focuses heavily on counterspace capabilities to deter aggression and threaten U.S. space assets. Most recently, Russian hackers conducted a cyberattack on the U.S. company Viasat because its communications satellites support Ukrainian defenses, in addition to carrying out attacks on SpaceX Starlink satellites.[41] The Russians have conducted several tests of direct-ascent ASAT missiles to destroy satellites,[42] DEWs to disrupt and disable satellites,[43] and co-orbital satellites that could be used to kinetically destroy hostile state satellites and spacecraft. These co-orbital satellites have also been used to conduct reconnaissance against U.S. satellites. The Russian Space Forces' ASAT capabilities are supplemented by the Russian Strategic Rocket Forces, which operate conventional and nuclear ballistic, cruise, and hypersonic missiles of all ranges with the potential to threaten terrestrial and space assets. The Russian Aerospace Forces also maintain the capability to attack satellites with air-launched ballistic missiles.[44] Russian intelligence services, the Main Directorate of the General Staff (GRU), and the Foreign Intelligence Service (SVR) provide cyber warfare and intelligence collection capabilities, threatening space assets by allowing Russia to disrupt and degrade capabilities via cyberattack or to develop countermeasures to U.S. assets with intelligence collected. Further, the Russian military maintains significant EW assets with proven capability of uplink and downlink jamming of satellites and satellite signals as well as spoofing GPS signals. The Russians maintain highly capable military counterspace programs able to directly threaten U.S. satellites, spacecraft, and ground support stations critical to the U.S. military and economy.

Russian Space Doctrine

The overarching goals of Russia's space program are to ensure Russian global leadership in space; to protect the country by supporting strategic military deterrence against any attacks; to monitor potentially hostile space objects and assets; to launch space assets at increasingly rapid rates; and to effectively operate space assets supporting terrestrial and space-based military, intelligence, and economic strategic goals. Russian space companies and government organizations struggled throughout the 2000s to achieve their goals, leading to significant doctrinal and organizational shifts. Between 2013 and 2016, the Russian civilian and military space programs went through several reorganizations after a number of failed space launches. The reorganizations led to greater government control and nationalization of key space companies. The Russian Federal Space Agency merged with the United Rocket and Space Corporation in January 2016, creating the current iteration of Roscosmos and the civilian space program. The Russians sought to reduce inefficiencies and failures that prevented their program from reaching and surpassing the United States. In August 2015, the Russian Air Force and the Russian Aerospace Defense Forces merged to create the Russian Aerospace Forces. The Russian Space Forces were created as a subordinate branch of the Russian Aerospace Forces during the reorganization to focus Russian military space efforts. The reorganizations have focused Russian efforts and directly support Russia's current 10-year plan and long-term strategic goals.

Moscow's long-term goals are directly affected by the Russian economy. Russia's economy is not as dependent on space infrastructure as are the economies of the United States, China, and Europe. However, the amount of money allocated to Russia's space program depends on the current state of the economy. In 2014, Russia's budget was equivalent to $5 billion, or 28 percent of NASA's budget in the same year. The budget then dropped to $1.6 billion in 2016 after oil prices declined and Western nations increased sanctions on Russia, causing the ruble to lose half its value.[45] In 2020, the Russian government approved a $2.3 billion budget.[46] Although the FKP-2025 originally estimated higher budgets, the actual budget is approved each year. The Russian space program is working to secure an additional $1.5 billion through 2022.[47] Sanctions on state-owned factories have hindered Russia's development of space infrastructure by restricting access to foreign-made electronics (the Russian invasion of Ukraine has pushed Western powers to impose additional sanctions, limiting Russian access to technology). Roscosmos is unable to make up for this loss because the program is struggling financially. For the past six years, Roscosmos has seen more net loss than net profit.[48] This makes the space program heavily reliant on government funding. Despite economic issues, the Russian government continues to press forward to achieve its long-term goals and will work to surpass the United States in space-based military, intelligence, communications, and economic assets.

NORTH KOREAN SPACE CAPABILITIES

The North Koreans outwardly view their space program as a vehicle for economic growth and technological development and as an expression of national pride. Inwardly, the program is likely seen as a method of improving strategic capabilities for national defense purposes. North Korean leadership regularly emphasizes the legitimate, peaceful nature of the space program in attempts to garner support. According to the Korean Central News Agency, in November 2021, a major space conference was held to discuss how the space program seeks to improve the North Korean economy and the livelihoods of the citizens of North Korea through peaceful means.[49] North Korea's civilian space agency, the National Aerospace Development Administration (NADA), is responsible for space development projects.[50] Although relatively unsuccessful, the North Koreans have built and operated SLVs and satellites. NADA launches SLVs from the Songhae and Tonghae Satellite Launching Stations, which it is upgrading despite promising to shut them down.[51]

North Korea does not have an overt military space program but could easily develop one using its civilian space and ballistic missile programs as a foundation. Despite having no overt connections, the civilian space program and the ballistic missile program are cooperating. NADA and several high-ranking officials have been sanctioned by the U.S. Treasury Department for connections to North Korea's ballistic missile program and for weapons proliferation.[52] Both programs are capable of threatening American assets in space and around the world. The North Koreans are openly vocal about their desires to improve their programs. The space program is planning on placing additional satellites into orbit, including attempts at geostationary orbits, significantly more difficult than their attempted insertions into low Earth orbit (LEO). These satellites are likely aimed at providing the military and government in general with additional national security capabilities.

The North Korean space program continues to progress despite consistent failures to achieve successful launches and satellite orbits. NADA operates several variants of the base model Unha SLV, which is capable of carrying small satellites of approximately 200 kilograms or smaller into LEO.[53] While the majority of SLV launches have failed, the North Koreans managed two successful satellite launches via the Unha. Both satellites failed shortly after being released and have never sent detectable signals back to Earth.[54] The satellites were overtly intended to provide North Korea with civilian Earth-observation capabilities but were likely test beds for military geospatial intelligence technologies.[55]

The North Koreans maintain ballistic missile, electronic warfare, and cyber warfare capabilities with the potential to threaten American space assets. Although the North Koreans have not directly tested the capability, North Korean intercontinental ballistic missiles (ICBMs) are theoretically capable of striking assets in space and deploying nuclear warheads into orbit to create EMPs in efforts to destroy and disable assets. The Korean People's Army (KPA) Missile Guidance Bureau operates the country's ICBMs and has recently tested several new missiles.[56] The KPA's Electronic Warfare Bureau directs EW activities, while the KPA's EW Jamming Regiment executes strategic jamming missions including uplink jamming against satellites, downlink jamming against ships and aircraft, and GPS spoofing.[57] The Reconnaissance General Bureau directs cyber warfare activities, while the Cyber Warfare Guidance Unit and associated advanced persistent threats (APTs) execute computer network operations against government, military, financial, telecommunications, and infrastructure targets. North Korean hackers—incredibly adept at stealing financial assets—have also demonstrated the ability to breach networks, secure footholds, conduct reconnaissance, and destroy data. North Korea has no publicly known directed energy or direct-ascent anti-satellite missile program but is likely working to develop such assets. The nation's missile, electronic warfare, and cyber warfare capabilities present a threat to space-based assets but are more likely to be used against ground control stations supporting those assets.

IRANIAN SPACE CAPABILITIES

Similar to North Korea, Iran is not in the same league as Russia and China with regard to major space power aspirations or capabilities. However, the nation is developing competencies that pose a threat to U.S. space assets. For Iran, a space program is a point of national pride, a demonstration of economic and technological prowess, and a cover through which to develop technologies critical for national security. While limited in size and capability, Iran's space program is growing and remains key to Iranian geopolitical objectives. The Iranian Supreme Space Council, which reports to the Supreme Leader, directs Iranian space policy and operations. Iran divides responsibilities for space operations between the civilian Iranian Space Agency (ISA), which was founded in 2003, and the military Islamic Revolutionary Guard Corps (IRGC) Aerospace Force Space Command, which was revealed to the public in 2020. The ISA coordinates "peaceful

activities . . . in the area of space science and technology," including designing and producing satellites and SLVs.[58] Further, the ISA's Iranian Space
Research Center (ISRC) and Astronautic Research Institute (ARI) are
responsible for developing new technologies for SLVs and satellites.[59] The
ISA, ISRC, and ARI have all been sanctioned by the U.S. Treasury Department for supporting Iran's ballistic missile program.[60] The ISA maintains
space launch facilities in Seman and Qom. While the organizations have
civilian application, they are providing direct cover for Iranian militarization of space and missile development.

The IRGC Space Command directly benefits from the ISA's activities;
for example, the IRGC's Qased SLV is likely a variant of the ISA's Simorgh
SLV.[61] The IRGC launched the Qased SLV from their space launch center in Shahroud. The IRGC Space Command is responsible for launching
military satellites into orbit to support IRGC operations. The IRGC leadership is far more vocal about the intentions of their program than the ISA.
The commander of the IRGC has extolled the importance of military space
capabilities for national security purposes and countering adversaries.[62]
The commander of IRGC Space Command has stated that "larger satellites
will be deployed at higher orbits" by the IRGC, indicating future militarization of space.[63] Further, the IRGC Aerospace Force deputy commander
has openly admitted that Iran's space launch program is used to advance
ballistic missile technologies.[64] The Iranian space program overall is in a
developmental stage focusing on research and development of space systems but with minimal success.

The Iranians have developed a limited domestic space launch capability and operate four SLVs that can carry small satellites into LEO. The
SLVs also present an opportunity for development of future ICBM and
ASAT missile programs. As of this writing, the ISA has completed four
successful civilian satellite launches, and the IRGC has completed one
military launch.[65] Aside from the Nour satellite, launched in April 2020,
Iranian satellites have broken down and reentered Earth's atmosphere
within months of launch. The Nour satellite's mission was to provide the
IRGC with geospatial intelligence, but it was ineffective in execution. For
example, Iran published two sets of imagery in 2020 targeting Al Udeid
Air Base in Doha, Qatar,[66] and oil tankers en route to Venezuela, and the
image quality was extremely poor.[67] The commander of the U.S. Space
Command confirmed that the satellite was rudimentary and ineffective at
providing actionable intelligence.[68] Although the Iranian space program
and capabilities are limited, Iran remains focused on improving them for
civilian and military use.

Iranian electronic, cyber, and directed energy assets complement Iran's
strategic deterrence and asymmetric warfare capabilities through directly
threatening U.S. space systems. The IRGC has demonstrated limited
uplink jamming of civilian satellites and has been accused of downlink

jamming and spoofing GPS signals.[69] The IRGC is developing new EW units and recently unveiled new EW systems.[70] The IRGC, Ministry of Intelligence and Security, and Iranian Advanced Persistent Threat (APT) Group cyber warfare capabilities have demonstrated the ability to conduct offensive cyber network operations against government, military, telecommunications, and critical infrastructure targets. Iranian hackers have breached networks, secured footholds, stolen and destroyed data, and crippled physical infrastructure. The Iranian directed energy program is less concerning, but Iran has been accused of utilizing lasers to disrupt U.S. satellite sensors.[71] Overall, Iranian threats to American space assets are limited but capable nonetheless.

MONITORING AND PROTECTING CISLUNAR SPACE

The security challenges presented to this point have been confined to the U.S. space architecture located in various Earth orbits and to space-related infrastructure on Earth. However, it is important not to neglect Cislunar space as an area that will be under contention. Cislunar can be defined as the volume of space that encompasses the Moon, the Moon's orbit, and the Moon's gravitational influence, but it is also commonly used to refer to the area beyond geostationary orbit. This section provides a visual aid of Cislunar space along with a description of the region, which is critical to understanding the potential threat to U.S. national security in the near and long terms.

To get a sense of scale, consider the distance of a satellite in a geosynchronous orbit (GEO) around Earth as a unit of measure—the distance from the center of Earth to the satellite is approximately 42,164 km, or 1 GEO.[72] Figure 3.2 demonstrates the size of Cislunar space using GEO radii as the unit of measurement to help gauge the long distance. The mass of objects in space and their respective distances affect orbits, so in Cislunar space, the large mass of Earth and the Moon both heavily influence the orbit of any spacecraft or satellite. In addition to portraying distance, the figure also shows Lagrange points L1–L5. A Center for Strategic and International Studies (CSIS) report defines the Lagrange points as "locations in space where the gravitational forces from two large bodies, like the Earth and the Moon, balance each other. Lagrange points are important because spacecraft or satellites can stay in a stable orbit without having to expend much fuel."[73] China, the European Space Agency, and NASA are among the countries and organizations that have taken advantage of the Lagrange points.[74] According to the CSIS report, "All systems of two large bodies contain five Lagrange points, two of which (L4 and L5) are stable, and three of which (L1, L2, and L3) are unstable, meaning spacecraft at those points require periodic corrections to maintain their position."[75] Understanding how gravitational effects impact orbits is critical

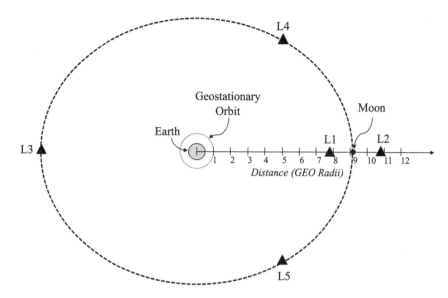

Figure 3.2   Spatial Scale in Cislunar Space

Source: The figure is based on Figure 1 in M. J. Holzinger, C. C. Chow, and P. Garretson, "A Primer on Cislunar Space," Air Force Research Laboratory, May 3, 2021, https://www.afrl.af.mil/Portals/90/Documents/RV/A%20Primer%20on%20Cislunar%20Space_Dist%20A_PA2021-1271.pdf?ver=vs6e0sE4PuJ51QC-15DEfg%3D%3D.

to grasping why the Lagrange points are strategically important. For example, a hostile actor could park an object at a Lagrange point to block others' access—there is a major strategic advantage for populating these quasi-stable orbits before an adversary does.[76]

Space sensors are located in various orbits around Earth and traditionally cover the near-Earth region, but it is imperative for the United States to shift its strategy to include an outward focus as well. New space sensor systems will be needed to track orbits outside of the traditional crafts orbiting Earth. The Earth-Moon system is not stationary, so the constant shifting of the two-body system necessitates multiple sensors near Lagrange points to minimize any coverage gaps in Cislunar space.[77] A senior intelligence officer with the Defense Intelligence Agency stated, "One of the things I always harp on in any forum: Our [Space Situational Awareness] is set up on an Earth-centric reference frame. . . . Once you get past a certain distance, it becomes very difficult to track things because the orbits don't repeat."[78] Defense officials have identified the launch of an ASAT weapon into a "translunar injection trajectory" as a significant area of concern. In that instance, an ASAT is launched as if it is going to the Moon, but once the ASAT reaches the Moon, it deceptively changes course into an orbit set to collide with a U.S.

national security satellite—effectively attacking the satellite from "above" as opposed to what we consider "below" in a more predictable direct launch against a target.[79] As such, Cislunar constitutes a "strategic flank" for U.S. space systems. Moreover, in the long term, the Moon itself may become a source for launch and manufacturing, able to introduce new platforms into space. The United States will need to drastically improve space situational awareness in Cislunar to ensure sensor systems are effectively tracking satellites outside the region near Earth so that hostile actors are not able to deny access to Lagrange points or the Moon or to provide attack opportunities to U.S. space assets.[80] Moreover, in the long term, regions around other planets and their moons may be strategically valuable areas to monitor and protect as competency in space improves and the United States continues to expand vulnerable space assets across the solar system.[81]

OVERVIEW OF CHALLENGES TO U.S. SPACE SECURITY

The United States faces broad space security challenges, including natural threats from solar storms, asteroids, and comets; anti-satellite threats to erode military advantage and hold the U.S. economy and critical infrastructure at risk; and strategic and coercive threats to America's vision of a vibrant space economy characterized by a liberal trading system rather than closed economic empires and resource nationalism. For the United States to prevail, it must have the necessary presence and force to shape and protect the domain as America's civil and commercial activities extend into deep space, it must have the resilience and counterspace capabilities to ensure the United States' advantage in terrestrial conflicts supporting its allies, and it must have the vigilance and in-space capabilities to protect the homeland from threats originating in space.

Chapter Highlights

- **Natural and passive space threats.** Solar storms from the Sun can send powerful bursts of charged particles at Earth, resulting in electromagnetic-pulse (EMP) weaponlike effects that interfere with communications and electrical infrastructure. The United States needs better mapping to identify asteroids and comets at risk of striking Earth. Small pieces of debris can prove lethal to satellites.

- **Anti-satellite (ASAT) weapons and kinetic and nonkinetic threats.** Counterspace threats can be classified into categories with varying destructive ability, including cyberspace and electronic warfare, directed energy weapons, kinetic energy attacks, and orbital threats. Attacks can target ground stations, communications links, and space infrastructure.

- **Chinese space capabilities and doctrine.** The PRC's space strategy focuses on accomplishing a set of long-term goals to compete with, and surpass, the United States by strategically emphasizing key areas that the United States typically neglects. PRC space-capable weapons systems include ASAT and ballistic missiles; co-orbital satellites; and directed energy, cyber, and electronic warfare capabilities.

- **Russian space capabilities and doctrine.** Russia's space program goals may be less comprehensive than China's, but it focuses heavily on counterspace capabilities to deter aggression and threaten U.S. space assets. Space funding hinges on the success or failure of oil and gas sales and current sanctions.

- **North Korean space capabilities and doctrine.** North Korea's space program is a method for improving strategic capabilities for national defense. North Korea maintains ballistic missile, electronic warfare, and cyber warfare capabilities to threaten U.S. space assets—but would likely attack ground control stations.

- **Iranian space capabilities and doctrine.** Iran's space program is in a developmental stage, focusing on research and development. Electronic, cyber, and directed energy assets complement strategic deterrence and asymmetric warfare capabilities by threatening space systems. Iranian lasers may have been used to disrupt U.S. satellites.

- **Monitoring and protecting Cislunar space.** Cislunar is the volume of space extending from Earth to the Moon, including the Moon's orbit and gravitational influence. Space sensor systems will be needed to track orbits outside of the traditional crafts orbiting Earth.

CHAPTER 4

American Space Primacy in Question

Richard M. Harrison, Peter A. Garretson, and Anthony Imperato

The U.S. space program was born out of political and national security motivations. When the Soviet Union launched Sputnik, the world's first satellite, on October 4, 1957, it garnered the attention of the international community and caught the American people by surprise.[1] The United States scrambled to respond, because, as President Lyndon B. Johnson famously stated, "Control of space means control of the world," and there were stark military implications of having the technical prowess to launch a satellite. Thus, a space race was born.[2]

Although the Soviet Union initially held the lead in that contest, the crewed Moon landing on July 20, 1969, marked an unparalleled technological achievement, solidifying America's place as the victor. For Americans growing up during the 1970s and 1980s, it was not uncommon to dream of becoming an astronaut and exploring the galaxy. It represented the dawn of a new era for fans of science fiction films and programs like *Star Wars* and *Star Trek*. People truly envisioned a future for humankind among the stars. And the dream seemed achievable. After all, two decades earlier, space military experts in the U.S. Air Force had projected that, within a decade, the Moon would play host to a military rocket base.[3]

Fast forward to 2022, however, and the picture is very different. Technological advancement in the space domain has not progressed nearly as forecast. We have not been able to get humans back to the Moon since 1972, nor have nations established long-term military installations there.[4] This slowdown is perhaps not surprising; the race to the Moon was driven by international prestige, and extreme levels of government funding followed. But after the United States' success in attaining that goal, there was no clear subsequent objective, and the excitement, pace, and focus of the American space effort simply proved unsustainable.[5]

DECLINING NATIONAL INTEREST AND INVESTMENT

Despite America's monumental head start in space a half century ago, the U.S. advantage relative to the rest of the world has eroded since. Indeed, until the successful May 2020 launch by SpaceX[6] proved the viability of commercial space launch options, NASA had been forced to rely on Russian rockets[7] to carry American astronauts to the International Space Station—a sad consequence of the shuttering of the Space Shuttle program in 2011.[8]

The United States' decline in the space domain can be attributed to a number of factors. These include a persistent lack of political will among U.S. policymaking elites, the absence of a clear and objective goal to accomplish, risk averseness and fear of failure, the substantial cost of space access, and government contractors who are disincentivized, by the bureaucratic and failed acquisition process from developing technology in a timely fashion.[9]

Yet polling suggests the U.S. general public still believes that space matters. Findings by the prestigious Pew Research Center indicate that over 70 percent of Americans believe the United States should remain a leader in space exploration. However, national space priorities don't currently match up with public sentiment; nearly 60 percent of those polled saw monitoring climate change or asteroids as a higher priority for NASA than going to the Moon or Mars.[10] The mismatch is more glaring still. At the time of writing, the U.S. vision for space does not include asteroid defense, climate change solutions such as space solar power, or the broader civilizational currents to settle space and become multiplanetary. Meanwhile, space has slowly but surely become the foundation of modern economic growth, societal discourse, and military power projection.

Despite the increased prominence of space and strong public interest, the United States is not adequately funding the space-related government agencies. Total U.S. space funding is not even at levels comparable to the Cold War as a percentage of gross domestic product (GDP) and the federal budget. A May 2021 American Foreign Policy Council report on the U.S. space budget outlined these funding deficiencies and discussed numerous alarming trends.[11]

First, as illustrated by figure 4.1, there has been a precipitous decline in space funding since the end of the Cold War. Today, total space funding amounts to less than 1 percent of the federal budget. During the Cold War, it was close to 2 percent.[12] Second, NASA funding has dipped below

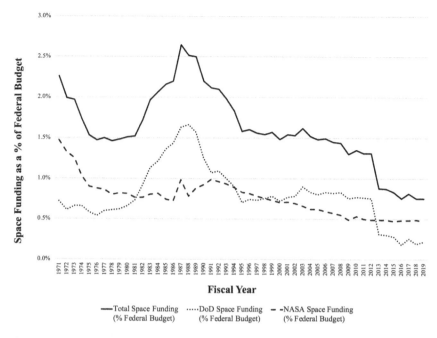

Figure 4.1   Space Funding Overview

Sources: "Aeronautics and Space Report of the President: Fiscal Year 2019 Activities," NASA, 2020, https://history.nasa.gov/presrep2019.pdf; "Federal Government Expenditures: Budget Outlays," Federal Reserve Economic Data, https://fred.stlouisfed.org/series/M318191A027NBEA.

half a percent of the federal budget, after remaining at close to 1 percent during the Cold War.[13] Third, sequestration has had a devastating effect on Department of Defense space funding—despite adding a new military branch (the U.S. Space Force). Throughout the Cold War, Defense Department space funding fluctuated between 1.0 and 1.5 percent of the federal budget and then held steady near 0.75 percent until the 2013 sequestration and has since averaged about 0.25 percent of the federal budget.[14] Moreover, in addition to the decline in space funding writ large, Defense Department science and technology funding, which spurs innovation, is a paltry 0.9 percent of the Space Force budget—significantly below the 3.4 percent recommended by Eric Schmidt, former chairman of the Pentagon's Defense Innovation Board.[15]

The figures are striking. When the United States was locked into a decades-long clash with the Soviet Union, space was viewed as a valuable investment. Yet today, at a time of renewed great power competition with China and increasingly accessible space technologies, the United States is not prioritizing space investment. At these levels, how can America continue to remain a leader in space? The conscious decision on the part of bureaucrats and politicians to reduce space funding runs counter to the boundless benefits of the burgeoning space economy. Space should not be thought of as a costly burden but, rather, as an investment consistently yielding solid returns.

A SPACE ECONOMY IS WORTH THE UP-FRONT EXPENSE

Over the past several decades, space has proven itself to be a viable resource, beneficial for both American economic prosperity and national security. The U.S. military and commercial space sector have been focused on developing numerous constellations of satellites with a host of applications ranging from weather to communications and imaging or surveillance gathering. In the national security space, the Pentagon benefits from space sensors that gather useful information such as providing precision positioning, navigation, and timing (PNT) services; intelligence, surveillance, and reconnaissance; detection of adversary missile launches and nuclear detonations; and military command, control, and communications.[16] On the economic front, nearly every advanced sector of society benefits from the use of U.S. space assets, including agriculture, aviation, the financial system, health care, telecommunications, and transportation.[17]

The U.S. investment in space has already paid major dividends for both the American economy and national security. Although NASA is not thought of as a revenue generator, in fiscal year (FY) 2019, the organization's enterprises contributed $64.3 billion in economic output (with about $14.2 billion coming from scientific research and development).[18] That massive figure amounts to an approximately 200 percent return on investment, when put in context with the $21.5 billion FY 2019 NASA budget.[19] Furthermore, in FY 2019, NASA provided 312,630 American jobs (48,912 of which were in the scientific research and development sector).[20] NASA missions and research have also provided many tangible benefits, such as technologies that have been readily adapted for everyday use on Earth—over 2,000 spin-off technologies were produced during the past 40 years.[21] Some examples of NASA-inspired technologies include advanced water filtration, digital image sensors, phase change materials, and advanced cardiac pumps.[22]

The Department of Defense has also significantly benefited the U.S. economy through its development of the Global Positioning System (GPS), which has cumulatively generated $1.4 trillion for the American economy since its inception several decades ago.[23] The GPS system, now operated by the U.S. Space Force, annually generates approximately $70 billion—close to five times the $15.4 billion FY 2021 Space Force budget.[24] Beyond revenue generation, the GPS satellite constellation also helps cut carbon emissions through improved navigation, resulting in an estimated 15–21 percent reduction in fuel expenditure.[25]

Advancements fostered by NASA missions and military-designed space assets have proven incredibly useful for the economy. However, the new major driver of a blossoming space economy is the private sector. Private-sector firms like SpaceX have made major advancements in technology with reusable launch vehicles and drastically reduced launch costs. Companies like Virgin Galactic and Blue Origin are beginning to make space tourism accessible to a broader audience. All told, prominent economists and banks now believe the space economy is poised to expand substantially from today's value of $450 billion annually to well over a trillion dollars per year in the next two decades (while China believes the space economy will be worth 10 times that figure). According to a Congressional Research Service report:

Global revenue from space-based services annually exceeds $300 billion, with more than two-thirds in the commercial sector. Well over $100 billion in annual revenues arises from commercial space data services (mostly direct-to-home television). Over $100 billion derives from commercial space equipment manufacturing. Finally, governments spend about $80 billion per year on space programs, with the U.S. government spending roughly 60% of that $80 billion.[26]

The study also cites a June 2015 report from the Department of Homeland Security that claims that approximately $1.6 trillion of annual U.S. business revenue is reliant on satellites.[27] As the space market continues to expand, it will become increasingly important to protect these valuable—and lucrative—space assets.

When considering the question of U.S. space primacy, it's important to zoom out and compare the United States with other nations. Table 4.1 portrays two critical data trends. First, it shows the vitality of the U.S. private sector, which understands the value of space and has significantly increased the rate of launches and amassed major constellations. Second, it shows the rapid advance of China's space program and its significant focus on military satellites.

INCREASING ADVERSARIAL COMPETITION

Amid declining U.S. funding for space development, our adversaries have not sat idly by. This is particularly true in the case of the People's Republic of China (PRC). The PRC sees space as a general enabler of its national power and as a way to improve its economic and military posture. From a commercial standpoint, PNT systems hold the power to drive banking and finance.[28] Various forms of space surveillance and imaging can help to develop or exploit natural resources as well as monitor weather

Table 4.1   Satellites Launched in Recent Decades

Total Satellites Launched				
Country	Type	2000–2009	2010–2019	2020–2021*
China	Military	1	102	33
China	Civilian government	22	127	67
China	Commercial	2	82	51
Russia	Military	24	69	8
Russia	Civilian government	1	21	5
Russia	Commercial	4	23	10
USA	Military	53	108	46
USA	Civilian government	15	140	28
USA	Commercial	62	390	2,019
World Total		353	1,569	2,847

* This column represents data from just two years, compared to the data reported for the preceding decades.

Sources: Information provided in the table was created using the UCS Satellite Database published on January 1, 2022, https://www.ucsusa.org/resources/satellite-database. Satellites listed in the database as "commercial/civil" are counted as civilian government, "government/commercial" are counted as civilian government, "military/commercial" are counted as military, and "military/government" are counted as military. This table excludes multinationally operated satellites.

and climate to support agricultural activity and military activity.[29] The Chinese use of technology to erode the security of other countries is well documented and is particularly prevalent in the cybersecurity sector and in the current debate over 5G technology.[30] Forms of surveillance can be used to support military operations or to warn of hostile military activity, and media and communications—bolstered by satellite systems—can more robustly promote culture and ideology.

The Chinese Communist Party understands this well. Back in 2016, the Central Committee and State Council issued a new National Innovation-Driven Development Strategy, which aspires to "vigorously improve technological capabilities for space entry and exploitation, perfect space infrastructure, promote technology development and application of satellite-based remote sensing, communications, navigation, positioning services, etc., and perfect the innovation chain for satellite applications."[31] China, moreover, is already using space as a component of its partnership-building diplomacy abroad[32] and as a prong of its Belt and Road Initiative.[33] It is engaged in aggressive space science diplomacy involving satellite design, construction, and launch, as well as advertised use of its space station. China's ambitions include not just space-based remote sensing and broadband (including 6G satellites), however.[34] China has set its sights on entirely new emerging markets, including space-based commercial and industrial facilities (2021+); space-based power generation (2030+); Lunar mining (2030+); asteroid mining (2034+); and, eventually, advanced space transportation.[35] These are all key components of a broader play for economic dominance. China has envisioned an Earth-Moon space economic zone producing $10 trillion a year by 2050, driven by key breakthroughs in the technologies outlined above.[36]

China's plan for space has much to do with its thirst for resources to fuel its ever-growing economy and reduce its reliance on other states for energy production. Three elements are essential for space dominance: space nuclear power, propellant, and space-based solar power. The PRC is now investing in all three to realize its long-term vision for space.

DEVELOPING A RENEWED INTEREST IN SPACE

In the past few years, the space sector has seen a marked resurgence in national security and private-sector interest. This has happened for several reasons:

- Private-sector space firms have reduced launch costs through innovative technologies.
- Because of the advent of hypersonic weapons, which require a space constellation of low-orbit satellites for tracking, there is renewed interest in space situational awareness for missile defense.[37]
- The executive branch has pushed for more funding in space and advocated for the creation of a new dedicated military branch for space.

Nevertheless, impediments remain. Unlike in China, U.S. space technology is advancing in silos rather than functioning in the concerted fashion necessary to achieve a competitive advantage. In other words, the United States still lacks a comprehensive vision for how to compete—and persevere—in the space domain. In an article for the *Space Review*, American Foreign Policy Council senior fellow for National Security Affairs Lamont Colucci summarized the situation this way:

Space policy is dominated by camps. One consists of the scientists who have little interest in the political-strategic equation and, in a few cases, work against it. A second camp is dominated by the traditional military, suspicious of ideas such as space domination and the need for a separate service. A third features the political class who may understand the immediate value of the space program, but fail to prioritize the right programs. A fourth and final camp includes some of the astronauts who see space exploration only in the context of exploration for exploration's sake. Rarely has anyone articulated where the USA needs to be in five, ten, fifty, or one hundred years—and beyond—to ensure it is the premiere spacefaring nation. The real priority is to fully integrate all these aspects of space into current and future national security and grand strategic thinking.[38]

Over the past decade, policymakers have advocated for space to varying degrees, but they have not done so in a truly comprehensive, coherent, and vigorous manner. The Obama administration published a national space policy,[39] but space was never a true priority for the administration aside from a focus on Earth science. Both the Obama and Trump administrations demonstrated an understanding of the need to protect assets in space (as exhibited by the Obama administration's National Security Space Strategy[40] and the Trump administration's Defense Space Strategy[41]). Until recently, however, both space agendas were limited in providing a comprehensive vision with concrete deliverables across space sectors.

To its credit, the Trump administration took several major steps to address America's inadequacy in space. Rather than unveil a comprehensive plan, the Trump administration resurrected the National Space Council[42] and opted to issue intermittent space policy directives addressing various key initiatives.[43] A notable space policy directive recognized the importance of space situational awareness (SSA) and space traffic management (STM), calling on multiple domestic space stakeholders to cooperate on making space safe from collisions and orbital debris. Its most controversial decision, however, was the formation of the Space Force as a new military branch and the concurrent reestablishment of a U.S. Space Command. In his February 2020 State of the Union address, President Trump also threw his weight behind NASA's Artemis program of crewed Lunar exploration.[44] And even after President Trump lost the 2020 election, his White House published a national space policy.[45] In much the same vein, the Trump White House's National Space Council continued releasing publications until the end of its time in office, including *A New Era for Deep Space Exploration and Development*, which outlined a persuasive rationale for the United States to focus on industrial development in space.[46]

The Biden administration has yet to make any major course correction from the previous administration's plans. It has opted to maintain the functioning of the National Space Council, and President Biden has continued to endorse both the Space Force and the Artemis program, even requesting a \$1.5 billion increase to the NASA budget.[47] This continuity and momentum needs to be sustained if the United States is to maintain its advantage in space.

GROWING THE U.S. SPACE ECONOMY

Encouraging, nurturing, and sustaining a thriving U.S. space economy is essential to ensuring U.S. economic and national security.

First, the United States needs to focus on developing cost-effective methods for the space program to prosper. For years, the United States has been plagued by major cost overruns and mired in delays by large defense contractors, which were not incentivized to innovate, meet performance expectations, and deliver space contracts on time. The NASA Space Launch System (SLS) rocket contract with primary contractors Boeing, Northrop Grumman, and Aerojet Rocketdyne is a textbook example of large contractor inefficiency. The project, which has been under development for a decade, was envisioned to serve as the largest heavy-launch vehicle that could launch humans and cargo to the Moon and Mars.[48] The SLS rocket has yet to be flown, but when it does, it will fly with an estimated cost of $4 *billion* per launch.[49] Compare that to the equally capable SpaceX Starship rocket, which is estimated to launch at $10 *million* per launch within the next three years (and which, unlike the SLS, will be fully reusable).[50] Clearly, the acquisition system is broken in its current state. A new methodology and incentives need to be in place to hasten the pace of progress in space and aid in growing the U.S. space economy.

Estimates for the current size of the domestic U.S. space economy vary, and there continues to be a widespread lack of data about the total size and scope of the sector. By some estimates, however, the U.S. space economy accounts for over half the global space economy.[51] In the December 2020 *Survey of Current Business*, the U.S. Bureau of Economic Analysis (BEA) unveiled preliminary estimates of the U.S. space economy for the period 2012–2018. The study calculated the value of the U.S. space economy as $177.5 billion in 2018.[52] It also accounted for over 356,000 private-sector jobs, highlighting an important feature. In addition to the plethora of economic benefits it has to offer, space provides an opportunity to invest in factories, skilled labor, and a transportation network.

Using these preliminary statistics from the BEA, the space advocacy organization Foundation for the Future developed two models to forecast the growth of the U.S. space economy from the present to 2050. Even under its most conservative model, the results were striking. The model assumes the space sector will grow by 4.5 percent annually and will maintain its workforce composition ratios between knowledge workers and skilled labor.[53] Based on those features, the model projects that the U.S. space economy will grow to $320.3 billion and will support 642,327 private-sector jobs by 2030. By 2040, these figures will increase to $461.2 billion and 925,088, respectively. And by 2050, the model projects that the U.S. space economy will have a value of $779.8 billion and will account for 1.6 million private-sector jobs.

Those numbers tell a clear story. The American space economy is at a critical juncture and is poised to grow rapidly. In this way, it is similar to two prior points in American history: westward expansion and the early days of the airline industry.

Westward Expansion

During America's expansion to the west, robust public-private partnerships helped to realize the transcontinental railroad, allowing a growing nation to truly extend its sovereignty from coast to coast. Indeed, President Abraham Lincoln was a major proponent of developing a transcontinental railroad system, and the 1860 Republican Party convention (which nominated him for the presidency) also adopted a party platform that called for the construction of the railroad system.[54] Lincoln believed that a transcontinental railroad system would allow for the easier and safer transportation of settlers traveling westward, promote national unity, and bolster interstate commerce and international trade.[55] Its construction was also prompted by national security concerns. Lincoln discerned that the railroad would enhance the United States' ability to defend the American frontier from foreign threats and aid the Union's military efforts against the Confederacy.[56]

Congress officially kick-started the construction of the transcontinental railroad with the Pacific Railroad Act, which passed on July 1, 1862.[57] To assist in the development of the railroad, the federal government was authorized to provide bonds for funding assistance as well as land grants.[58] Construction officially began in 1863 and was completed in 1869, linking the United States from coast to coast.[59] The United States was the first nation to construct a transcontinental railway system.[60] In turn, the accomplishment allowed for expanded interstate travel and commerce, promoted U.S. global trade, and spurred the industrial revolution in the succeeding decades.[61]

The results were nothing short of revolutionary. As depicted in figure 4.2, the construction of the transcontinental railroad fueled a growth in the U.S. percentage of global GDP. By the end of the nineteenth century, the United States had cemented itself as a major global economic power.

The construction of the transcontinental railroad is analogous to the development of space logistics infrastructure that will be essential for the United States in its quest to maintain its status as the dominant space power and to benefit from the potential of the space economy. To achieve these benefits, the United States must work to expand public-private partnerships and make robust investments in the key space sectors (such as reusable rocket development; space tourism and services; space-based solar energy; nuclear power and propulsion; asteroid and Lunar mining; in-space servicing, assembly, and manufacturing; and Lunar economic development). Sustaining U.S. leadership in space will require achieving first-mover advantage in the development of space logistics infrastructure. And with the Chinese Communist Party working diligently to supplant the United States as the hegemonic space power, time is of essence.

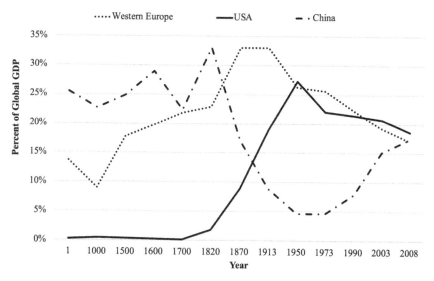

Figure 4.2   Shares of Global GDP

Source: The chart was created using information from the Maddison Project Database 2020. Maddison Project Database, version 2020. Bolt, Jutta, and Jan Luiten van Zanden (2020), "Maddison style estimates of the evolution of the world economy. A new 2020 update," https://www.rug.nl/ggdc/historicaldevelopment/maddison /releases/maddison-project-database-2020.

The Rise of Aviation

The rise of the aviation industry in the United States likewise offers a useful historical analogy to conceptualize the importance of increasing investments in critical space sectors and developing public-private part-nerships in the space domain. The U.S. aviation sector officially emerged on December 17, 1903, when Orville and Wilbur Wright conducted four short flights with their engineered aircraft, marking the start of aviation in the United States.[62] A little over a decade later, the St. Petersburg-Tampa Airboat Line, the first commercial airline service, conducted its inaugural flight.[63] Although the airline would last only four months, it marked the beginning of commercial air transportation and laid the foundation for present-day global air travel.[64] And while aviation was slow to take off in the United States in the early 1900s, America's par-ticipation in World War I changed everything.[65] On entering the war, the United States took action to mobilize the industrial sector to produce air-craft, relying mainly on the automobile sector.[66] Following World War I, in 1918, the civil aviation industry began to emerge in the United States, most conspicuously via the launch of airmail service between New York City and Washington, DC.[67] By 1920, the United States had established a transcontinental airmail route linking New York City and San Francisco.[68] Additional airmail branch lines were created subsequently, linking other American cities.[69] Over time, Congress opened the airmail system to private airlines.

With its entry into World War II, the United States mobilized industry to ramp up production of military equipment and other wartime necessities. U.S. aircraft production increased dramatically as well, as factories were repurposed to produce such equipment.[70] U.S. industrial mobilization and airpower played a key role in the Allied victory, with the United States acting as the "arsenal of democracy."[71] By the end of the war, the United States had manufactured over 300,000 military aircraft (over one-third of all military aircraft manufactured during the war, dwarfing all other nations).[72] After World War II, demand for commercial air travel began to surge, leading to the creation of new airlines and aviation companies, as well as further technological innovations.[73] And as the industry began to mature and prosper, the federal government was able to wind down its provision of subsidies to airlines.[74]

The efforts of both the public and private sectors led the United States to develop a robust aviation industry. The industry continues to play a major role in the U.S. economy, employing around 1.3 million Americans and accounting for 5.2 percent of total U.S. economic output.[75] It also facilitates both international trade and domestic commerce and makes important contributions to America's national defense.

The scale and scope of the economic possibilities in space are clear. Both the transcontinental railroad and the aviation industry demonstrate that, by developing effective public-private partnerships, the nation can greatly benefit from all that the space economy has to offer. Although the same rate of economic expansion is not guaranteed for the space industry, the United States nonetheless needs to position itself to lead the potential economic revolution based on an as-yet untapped market in the realm of space. This, however, will require economic and industrial output and innovation that will fundamentally change the international economic system in ways not seen since the industrial revolution.

CONTEMPLATING A U.S. SPACE STRATEGY

The lack of a strategy or a unified American space vision should not be surprising, since the relevant space-related government agencies each have their own, sometimes competing, interests, which also differ from those of companies in the private sector. Adding to the problem, multiple government agencies are involved in space to varying degrees, and their approaches will invariably need to be coordinated to contribute effectively to American success in space.[76] Despite this complexity, however, the United States still requires a whole-of-nation strategy for space—one that spans multiple presidential administrations.

Congress, to its credit, understands the need to address the unfolding competition in space. In the National Defense Authorization Act for Fiscal Year 2021 (Public Law 116–283), Congress specifically called for a space strategy to compete with China, instructing the National Space Council to "submit to the appropriate congressional committees an interagency assessment of the ability of the United States to compete with the space programs of China."[77] Elements of the law included a series of items that should be completed by July 2022:[78]

(A) A comprehensive assessment between the United States and China on—
  (i) human exploration and spaceflight capabilities;
  (ii) the viability and potential environmental impacts of extraction of space-based precious minerals, onsite exploitation of space-based natural resources, and the use of space-based solar power;
  (iii) the strategic interest in and capabilities for Cislunar space; and
  (iv) current and future space launch capabilities.
(B) The extent of foreign investment in the commercial space sector ...
(C) An assessment of the ability, role, costs, and authorities of the Department of Defense to mitigate the threats to commercial communications and navigation in space from the growing counterspace capabilities of China.
(D) An assessment of how China's activities are impacting the national security of the United States with respect to space, including—
  (i) theft of U.S. intellectual property; and
  (ii) efforts by China to seize control of critical elements of the U.S. space industrial supply chain and U.S. space industry companies.
(E) An assessment of efforts by China to pursue cooperative agreements with other nations to advance space development.
(F) Recommendations to Congress, including recommendations with respect to any legislative proposals to address threats by China to U.S. national space programs and the domestic commercial launch and satellite industries.

As of this writing, the reports requested by Congress have yet to be finalized, further demonstrating the United States' inability to address the Chinese space threat in a timely manner. Yet, even though exploring competition is essential, it is not the only reason to focus on space. Space experts have a wide array of answers regarding why space is significant and offer a range of purposes for going there. Here are just a few examples of views from across the spectrum, excerpted from the authors' *Space Strategy* podcast:

Argument 1: Expanding Human Life Throughout the Universe Is Essential.

We are in the most important 100 years of human history, and we are in the last decade or two of that 100 years; the decisions we make right now will determine whether the human species and life itself will expand beyond the planet Earth or the human species and life itself will be gone. ... I believe in my core that the purpose of humanity is to spread beyond the earth and plant the seeds of life in places that are dead.[79]

—Rick Tumlinson, founder
of SpaceFund

Argument 2: Increasing Space Power Translates to National Power.

Because anything that you can do in space, if you think about it hard enough, will probably be able to give you more national power throughout all the instruments of power that we have in the country. I have a very expansive view of what space power is, but it's a subset of national power, which means, if you are doing space correctly, what you do in space helps you become a stronger nation, a stronger country.[80]

—Brent Ziarnick, professor at the Air Command
and Staff College

Argument 3: Space Offers Limitless Opportunity.

The reason to go into space is not out of fear of extinction. The reason to go into space is out of excitement for a fantastic future for us and for our species. ... And the exciting near-term potential is, there is gold in them hills—there are fortunes to be made.[81]

—Joel Sercel, founder and CEO of
TransAstra Corporation

Argument 4: Alleviate Environmental Stress and Strain.

We're a growing species, and I really do think that we're starting to feel the stress and strain of population on a limited ... finite surface of the Earth. ... We can't ignore space, because it will be exploited and used against us, if we allow that to happen.[82]

—Gen. Steven Butow, director of the space portfolio
at the Defense Innovation Unit

Argument 5: Space Is a Domain for Commerce.

The value proposition for space has dramatically changed over the last 20 years. ... [W]e are in fact going to move from a discovery architecture in space to a commerce architecture in space.[83]

—Gen. James "Hoss" Cartwright, former vice chairman of the U.S. Joint Chiefs of Staff

Argument 6: Cultivating a New Generation of Scientists and Engineers.

An expansive space program would tell every young person a message— learn your science and you can become part of the great adventure, you can be a pioneer of new worlds. ... We would get tremendous intellectual capital, which would benefit us in every field. I myself, of course, went into science, because of the Apollo program.[84]

—Dr. Robert Zubrin, founder of the Mars Society

This wide array of views demonstrates the vast interest in space that currently exists and highlights the myriad reasons to go there. However, while each of these arguments has merit, there is a need for a consolidated strategy that can execute and prioritize them to ensure the United States is well positioned to be the world's dominant space power.

Fundamentally, the United States needs to go to space today to ensure protection of its space economic assets, to ensure the spread of humanity to the stars, and to continue its dominance in the global system. In doing so, its people will benefit from new opportunities in a burgeoning field, it will reduce strain on world ecology by moving industry off-world, and it will provide a unifying goal for the United States to pursue. To do so, however, the United States will need to own the largest share of the space economy, be the most significant space actor and leader, be the space actor with the initiative and primary rulemaking ability, and be perceived as the most powerful military when untested and as having the ability to prevail if tested. Without the political will and a defined approach, the United States may be surpassed by China on these fronts.

Chapter Highlights

- **Declining national interest and investment.** The national decline in space can be attributed to persistent lack of political will; absence of a clear goal to accomplish; risk averseness and fear of failure; the substantial cost of space access; and government contractors being disincentivized from developing technology quickly. Total space funding is less than 1 percent of the federal budget, compared to approximately 2 percent during the Cold War.

- **A space economy is worth the up-front expense.** U.S. investment in space pays major dividends. GPS, now operated by the U.S. Space Force, annually generates about $70 billion, five times the FY 2021 Space Force budget. NASA technology has produced 2,000 spin-off technologies in the past 40 years. The space economy is poised to be worth over a trillion dollars annually within two decades.

- **Increasing adversarial competition.** The PRC sees space as a general enabler of its national power and as a way to improve its economic and military posture. Three elements are essential for space dominance: space nuclear power, propellant, and space-based solar power. China is now investing in all three.

- **Developing a renewed interest in space.** The space sector has seen a resurgence because private-sector firms have reduced launch costs; hypersonic weapons, which require a space constellation of low-orbit satellites for tracking, have renewed interest in space situational awareness for missile defense; and the executive branch has pushed for more funding and advocated for the creation of a new military branch for space.

- **Growing the U.S. space economy.** By 2040, the U.S. space economy will grow to $460 billion and support 925,000 private-sector jobs. The historical examples of the transcontinental railroad and the aviation industry demonstrate that by developing effective public-private partnerships, the United States can benefit from the potential of the space economy.

- **Contemplating a U.S. space strategy.** Space experts suggest pursuing space to expand human life, increase space and national power, make money, alleviate stresses on the environment, increase commerce, and cultivate a new generation of scientists and engineers. There is a need for a consolidated strategy that can execute and prioritize these points to ensure the United States is well positioned to be the dominant space power.

CHAPTER 5

The Future of the U.S. Space Force

Richard M. Harrison and Peter A. Garretson

The new strategic challenges posed by space have made it necessary to alter both military organization and doctrine. By guaranteeing the security of American space assets, the U.S. government can further incentivize commercial investment in space. This requires the United States to have a robust security presence in that domain—as well as a clear understanding of authorities and priorities for the new U.S. Space Force (USSF) and U.S. Space Command and how both can complement the other U.S. military services. Yet the creation of the USSF was initially met with considerable ridicule, and its potential effectiveness remains poorly understood in policy circles even today.

Here we ask: What is the future of the USSF? With the long view of comprehensive national power and America's vision of in-space industrial power and spacefaring in mind, what will be asked of the USSF? What should be asked of the USSF in the present to ensure a vibrant future for America on the final frontier? With the vast changes taking place within the space domain, how will the activities and responsibilities of USSPACE-COM change, and how will those responsibilities drive the USSF's focus and composition?

RATIONALE FOR A NEW MILITARY SERVICE

In the eyes of the public, the push for a space force appeared to emerge suddenly from the initiative of President Donald Trump. In reality, however, the need for an independent space force had long been debated within the military community, almost since the dawn of the space age itself, reflecting a bipartisan congressional consensus that space was becoming such a central axis of military and strategic advantage that it needed to be managed centrally to allow direct strategic resourcing, shaping, and management by Congress.

The U.S. military generally organizes itself by domain—land, sea, air—and by the corresponding services—Army, Navy, Air Force. This is because the physics that dominate the domains result in very different sorts of platforms—tanks, ships, aircraft—to move and fight efficiently in those domains. In addition, the challenges of positioning, coordinating, and fighting result in very different styles of warfare, which necessitates distinct expertise that takes time to develop. Space was unique in that it was both a separate physical medium and a separate legal domain, yet it had no service dedicated to excellence in spacecraft or developing domain-specific expertise. Military space advocates long argued that for the same reasons it was desirable to separate the U.S. Air Force from the Army, it was desirable to separate a space force from the Air Force.

Prior to the establishment of the USSF, responsibilities for organizing, training, equipping, and operating military space forces were broadly distributed among a large number of stakeholder organizations, largely operating without a unified management chain providing strategic direction. The U.S. Army, Navy, and Air Force all had space capabilities, with the largest share by far—in excess of 90 percent—residing in the Air Force under the Air Force Space Command. The first U.S. Space Command operated from 1985 to 2002 but was deactivated in the wake of the September 11, 2001, attacks as the military switched focus to the global war on terrorism.

Detractors thought the USSF was not needed, since they erroneously believed space assets were needed only in support of ground operations (such as counterterrorism) and only to supply force enhancement services such as navigation and communication.[1] As such, they argued that the USSF would pull resources away from the Air Force, further militarize space, and encourage development of expensive new weapons systems.[2] Conversely, advocates argued succinctly that a U.S. Space Force was necessary "to consolidate authority and responsibility for national security space in a single chain of command; to build a robust cadre of space professionals who can develop space-centric strategy and doctrine; and to avoid the conflicts of interest inherent in the other Services that have shortchanged space programs for decades."[3]

On December 20, 2019, the United States took the dramatic step of creating the U.S. Space Force as a separate military service, reassigning approximately 16,000 military and civilian space personnel to the new branch (USSF personnel are termed Guardians).[4] Military services are powerful, persistent organizations established in law that organize, train, and equip forces for their medium. They establish the identity of their members and create cultures that advance or constrain strategic options.

The United States also reestablished the U.S. Space Command (USSPACE-COM), a combatant command (COCOM). COCOMs are directly responsible to the president to employ forces in military operations. COCOMs are also important organizations, since they employ forces, make plans for conflict, and establish requirements for the services. COCOMs are generally weaker bureaucratic actors because all their personnel are on temporary assignment from the services that control their careers, their responsibilities are established by the president's Unified Command Plan rather than Congress, they are customers rather than suppliers, and they have much smaller budgets than the services (e.g., the USSPACECOM budget is barely over 1 percent of the USSF budget).[5] Nevertheless, as long as America chooses to split the unity of command between services and COCOMs, a question about the future of the USSF is simultaneously a question about the future of USSPACECOM.

While it was the Trump administration that received the most attention for establishing the USSF and USSPACECOM, several lines of concern led Congress to conclude that legislation was required to secure American advantage in space:

- The rapid progress of other nations in testing and procuring military systems, compared to the slow rate of innovation in U.S. military space systems and poor performance in acquisition was a clear symptom that the United States was not optimally organized to maintain advantage in space.
- The failure of the U.S. Air Force to provide a separate career path to develop space expertise and the lack of progress in space doctrine suggested to Congress that subordination of space professionals within the Air Force was preventing progress.[6]
- The large and complicated number of stakeholders in the space domain and subordination of space budgets to the Air Force frustrated the efforts of Congress to get straight answers and feel that someone was directly accountable for success.[7]
- Congress was frustrated by the failures to standardize space systems and ground user equipment across the services.
- The steady progress of Chinese and Russian anti-satellite (ASAT) systems and the persistent vulnerability of U.S. systems was a symptom that the problem was acute and immediate enough to require action.[8] As members of Congress became more aware of the threats in classified briefings,[9] the status quo and slow pace of progress became untenable.

These concerns tipped Congress to provide an organizational fix whereby a single organization could focus on the development of military space systems, doctrine, and professionals as well as to centralize accountability.

But there were wider concerns as well. Both in the administration and Congress there was an awareness that, as a result of state initiatives and commercial innovation, space was transitioning from a domain of discovery to a domain of commerce and that, to secure American long-term strategic and economic interests, there must be a uniformed force—analogous to the U.S. Navy or Coast Guard on the high seas—to provide security for commerce.[10] The government began to recognize the overdue need to embark on a peacetime strategic offensive to gain the advantage for strategic ends beyond and antecedent to war-fighting in the domain itself. U.S. naval officer and historian Alfred Mahan wrote, "Naval Strategy has for its end to found, support, and increase, as well in peace as in war, the sea power of a country."[11] Space strategy in USSPACECOM should expand spacepower throughout the competition continuum.[12] Further, there was a need to develop a cadre of military professionals who looked at the domain as a theater itself, not only as a supporting theater, but as a primary theater for strategic competition. Because of the long lead time to create such capabilities to economize this new strategic frontier, it was felt that early investment was necessary to be ready for a projected future in which there were commercial interests and economic activity in space, including on the Moon and asteroids.

USSF PRIORITIES

Two foundational documents laid out the justification and role for the USSF (the President's Space Policy Directive-4)[13] and formally established the USSF and its enumerated missions (the National Defense Authorization Act for Fiscal Year 2020).[14] In August 2020, the USSF unveiled "Spacepower: Doctrine for Space Forces,"[15] which articulated the USSF vision, guiding principles for military spacepower, and the importance of spacepower in safeguarding America's national security interests. However, these documents did not provide sufficient guidance for the development of personnel, culture, and systems.

How clearly defined is the role and mission of the USSF? According to the official USSF website, the USSF "is a military service that organizes, trains, and equips space forces in order to protect U.S. and allied interests in space and to provide space capabilities to the joint force. USSF responsibilities will include developing Guardians, acquiring military space systems, maturing the military doctrine for spacepower, and organizing space forces to present to our Combatant Commands."[16] The mission description is not comprehensive, which may have been purposeful to allow for positive mission creep, or because the necessary responsibilities are as yet unknown.

Before exploring the various priorities of the USSF in detail, it is important to understand the service's area of responsibility because that can bound the scope of mission. In "Space Operations" (*Joint Publication 3–14*), a section of text delineates the geographic area of responsibility (AOR) for SPACECOM and, correspondingly, that of the USSF:

SPACECOM's AOR is the area surrounding the Earth at altitudes equal to, or greater than, 100 kilometers (54 nautical miles) above mean sea level. The relationship between space and cyberspace is unique in that many space operations depend on cyberspace, and a critical portion of cyberspace can only be provided via space operations. . . . Through effective coordination of AOR boundaries, combatant commanders (CCDRs) help to preserve space situational awareness (SSA), spacecraft life span, and space system performance. This, in turn, facilitates freedom of action in space and improves space support to terrestrial operations. It is important to note that missile defense operations transiting through the AOR are not pre-coordinated due to the short-/no-notice self-defense actions required to defeat enemy ballistic missile attacks.[17]

Thus, the defined operational area for the USSF is effectively about 100 km from Earth's surface and stretching to eternity. This defined space outside of the traditional military global zone where conflicts occur may be termed *supraglobal*, to refer to "those things that are relevant to military or political matters that encompass the globe and relevant activities in the space beyond it."[18]

Air force space operations were previously limited to traditional low, medium, highly elliptical, and geosynchronous Earth orbits, but during the 2010s, the space mission AOR began to broaden as the United States and other countries planned commercial projects, trans-Lunar injection trajectories, and operations at Lagrange points.[19] With such an enormous amount of territory to consider, it is imperative to prioritize specific mission sets for the new service. The Space Force will be organized, trained, and equipped to meet the following priorities:

- Protecting the nation's interests in space and the peaceful use of space for all responsible actors, consistent with applicable law, including international law;
- Ensuring unfettered use of space for U.S. national security purposes; the U.S. economy; and U.S. persons, partners, and allies;
- Deterring aggression and defending the nation, U.S. allies, and U.S. interests from hostile acts in and from space;
- Ensuring that needed space capabilities are integrated and available to all U.S. Combatant Commands;
- Projecting military power in, from, and to space in support of our nation's interests; and
- Developing, maintaining, and improving a community of professionals focused on the national security demands of the space domain.[20]

Upon creation of the USSF, the service's established role did not depart significantly from the role of space-focused personnel in the Air Force. In its current form, the military service is meant to have a more directed focus on

space situational awareness; satellite operations and global, integrated, command and control of military space forces; global and theater military space operations to enable joint campaigns (to include missile warning); space support to land, air, naval, and cyber forces; spacelift and space range operations; space-based nuclear detonation detection; and prompt and sustained offensive and defensive space operations to achieve space superiority.[21]

The fundamental change was the realization that the existing organizational structure (the Air Force Space Command) was incapable of taking advantage of the vast opportunities and strategic interests in space and that only an organizational movement toward independence could prevent the United States from losing ground on its interests in a key strategic theater, space itself. However, the USSF should not be focused solely on protection of satellites and with only military objectives in mind. Rather, the USSF should be focused on accumulating spacepower for the nation. The USSF budget shows where the priorities are, with recent increases in missile warning and tracking and expected increases for tactical intelligence, surveillance, and reconnaissance (ISR) starting with ground moving target indicators in next year's budget.

STRIVING FOR SPACEPOWER

There is no universally accepted definition of spacepower, but in the most simple terms, accruing spacepower allows a country to better accomplish its objectives through space activity.[22] Though the definition is exceedingly broad, holding spacepower is an incredibly valuable asset. According to space strategist Brent D. Ziarnick, "[T]he sole utility of spacepower to the state is to expand the general power available to the state through space activity. Space activity can enhance the power of the state across all instruments of national power: diplomatic, informational, military, and economic (DIME)."[23] Ziarnick provides a relatively easy way to think about spacepower as controlling the space domain to ensure a nation's freedom of operation and its ability to generate economic returns and deny the ability for an adversary to gain an economic or military advantage. The USSF may not have been conceived to maximize American spacepower, but if the United States aspires to become the preeminent space power, the USSF will play a critical role. Along those same lines, influential space advocate and entrepreneur Rick Tumlinson highlights the importance of spacepower and the new service, saying that "the Space Force is designed to solve the problem of how the

United States Government can best take advantage of human activity in space."[24]

Maintaining spacepower will require having the military might to foster and protect commercial development in space and continuously generate wealth-producing national power. This can be viewed as a virtuous cycle for space, as commercial endeavors cultivate the ability to afford military superiority and provide increasing access to the infinite resources space has to offer (see figure 5.1).[25]

However, positioning the USSF to help America achieve the previously described vision and level of spacepower will take significant time and also require policymaker buy-in. After 70 years without any changes to the armed forces, a new strategic challenge necessitates shifts in military

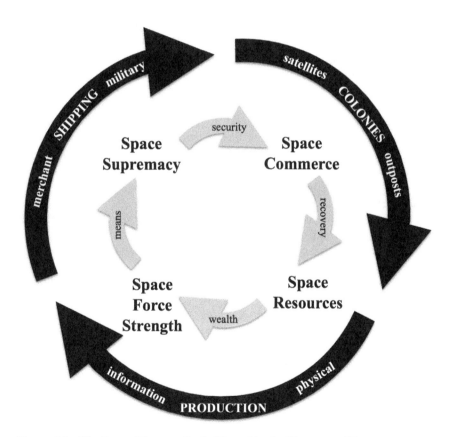

Figure 5.1 The Space Virtuous Cycle Nested in the Elements of Spacepower

Source: Brent Ziarnick, "A Practical Guide for Spacepower Strategy," *Space Force Journal,* Issue 1, January 31, 2021, https://spaceforcejournal.org/a-practical-guide -for-spacepower-strategy/.

organization and doctrine. The creation of the USSF does not yet ensure that the proper focus and resources will be given to providing for security and economic prosperity in the space domain, but it is a big major step forward for America.

Continued investment in the USSF is crucial for commercial opportunities to flourish in space and for investors to have confidence in the emerging space market. As Mir Sadat, former National Security Council policy director on space, has explained: "It will be the U.S. Space Force which will provide the necessary expertise for the U.S. Space Command to ensure unfettered access to and the freedom to operate within space[,] . . . just as . . . the U.S. Navy stands watch to ensure that we freely navigate the world's seas."[26]

Comparing future space operations with those in the maritime arena offers a valid analogy. In the nineteenth century, industrial age powers had to build global networks of infrastructure to sustain the projection of naval power in defense of their expanding commercial interests. The same can be said today for space. The USSF will serve to permit U.S. commerce to flow and flourish in space much the same way the U.S. Navy has historically guarded freedom of navigation on the high seas.[27] The current and future role of the USSF is still in question, and there is a debate about how the USSF should be operating with its role in the space domain. Comparisons have been made to classifications in the maritime domain as "brown-water" (sailing locally and supporting conflicts near land) versus "blue-water" (sailing in the open ocean and providing support to commercial activities). These classifications are analogous to the USSF operating in space near Earth and the Moon or in deeper Cislunar space and beyond.[28] Irrespective of the terms used to describe the operational environment, the reality is that the USSF will be conducting missions in an ever-expanding set of circumstances as the U.S. government moves to increase spacepower.

Policymakers have not yet been given the guidance needed to structure the USSF to properly address threats and play a defined role in ensuring U.S. space primacy. Additional work is necessary on this front, some of which has been outlined in proposals to relevant policy circles (e.g., Congress).[29] But questions remain, chief among them the appropriate roles and objectives of departmental agencies in ensuring national primacy in a key area of great power competition. Also, a discussion will need to follow regarding congressional input on offensive and defensive measures in space.

SPACE MILITARY OPERATIONS

To compete successfully against the strategic threat posed by China, the United States in general and the USSF in particular will need to be put on the strategic offensive at peace as aggressively as at war. The strategic threat posed by China in space is as yet grossly underappreciated. As a result, the United States might be said to be on the defensive, both strategically (since it lacks a plan to occupy and develop positions of strategic value in space) and tactically (because it has not placed major emphasis on developing offensive capabilities that it might deploy to deny others advantage).

USSF military planners need to be aware of nontraditional approaches to space warfare. Strategists may tend to focus on threats emanating from the ground such as ASAT systems or satellite-borne threats in traditional Earth orbits. However, there are threats from the whole Cislunar space theater that will need to be addressed. For example, China may decide to deny access to strategic locations on the Moon while pursuing helium-3 for nuclear fusion technology or potentially stationing intelligence-gathering satellites and weapons platforms at Lagrange points (see chapter 3 for a discussion of Cislunar monitoring and protection).[30] The counterspace capability that the USSF develops is important to ensure that space support is available to U.S. forces and that there are protective measures for U.S. assets in space.[31]

To date, the United States has been hesitant to demonstrate USSF military capabilities. However, the United States is capable of launching several types of counterspace capabilities, of varying destructive ability, on both space and ground infrastructure, as outlined in chapter 3. In a comprehensive open source assessment conducted by the Secure World Foundation, experts determined the United States has expertise in "counterspace development across co-orbital, direct ascent, directed energy, electronic warfare, and space situational awareness categories."[32] However, because most of the space weapons development occurs in the classified domain, little information about the extent of U.S. military capabilities is available publicly.[33]

The Counter Communications System Block 10.2 is one of the few systems that has been openly discussed by the USSF. Details are scarce regarding its effectiveness, but the system has the reported ability to disrupt or degrade adversary communications satellites through a jamming technique, preventing them from communicating effectively during a conflict.[34]

Though USSF weapons capabilities are not readily apparent, according to the spacepower doctrine, the USSF trains to "outwit, outmaneuver, and dominate thinking, competent, and lethal aggressors who are attempting to thwart U.S. actions."[35] Further the established doctrine provides a discussion of the importance of having an influential impact on the physical, cognitive, and network dimensions of the space domain.[36] The cybersecurity component, which is encapsulated in the network dimension, is of particular importance: "[S]ome examples of tactical maneuvers in this dimension include monitoring and defending software, attacking adversary computer systems, duplicating networks, amplifying signals, shifting frequencies, upgrading encryption, and adjusting data pathways. In a military context, links and nodes that comprise the network dimension of the space domain are potential vulnerabilities and subject to attack."[37] To date, however, the military has relied on a posture of deterrence in space, rather than a show of force to demonstrate capabilities.

Deterrence in Space

The incredibly capable U.S. military is constructed with the hope of never needing to go to war, but with the ability to persevere with overwhelming force if necessary. In the domains of air, land, and sea, the respective military branches have demonstrated their war-fighting capabilities on numerous occasions; however, in the space domain, the U.S. combat ability is somewhat opaque. The USSF has relied on a posture of deterrence, "the prevention of action by the existence of a credible threat of unacceptable counteraction and/or the belief that the cost of action outweighs the perceived benefit,"[38] to keep adversarial nations from attacking U.S. space assets. The United States reserves the right to respond in the time, place, and domain of choice, through multidomain operations or integrated deterrence.

Unfortunately, some military leaders believe that deterrence may not be enough.[39] As of this writing, there have been no major conflicts with a space warfare–capable adversary. But should a conflict occur, enemies may not be dissuaded from striking U.S. satellites—particularly GPS satellites operated by the military but relied on for countless civilian applications. China has already dazzled U.S. satellites, and Russia blew apart one of its own satellites with an ASAT test—and both nations have conducted close approaches to U.S. satellites during peacetime. These incidents do not lend credence to the likelihood that the nations will be deterred from space aggression in times of war.

Furthermore, a common concern is that the U.S. military has overly classified its capabilities, causing former vice chairman of the Joint Chiefs Gen. John E. Hyten and others to point out that you cannot deter an adversary with a weapon they have never seen. At an event with the National Security Space Association, Hyten said, "[D]eterrence does not happen in the classified world. Deterrence does not happen in the black; deterrence happens in the white."[40] One Department of Defense official argued that deterrence should be considered during weapon development, saying, "Did you conceive of the capability with the idea that you would reveal it? Because if you didn't, you shouldn't be revealing it now, or you should really think hard before revealing it. We need to design things that can be revealed without eliminating their effectiveness, and without causing escalation."[41] And not all adversaries can be deterred in the same manner—for example, China and Russia have different governing and military structures and distinct economic ties with the United States, so they are likely to respond to an American deterrence strategy in very different ways.[42] There are ongoing strategic reviews evaluating the mix of offensive and defensive systems and declassifying some.

In addition to countering any system that is used to attack the U.S. space architecture, it is also important to demonstrate the ability to conduct strikes on adversary space assets both in orbit and on ground-based infrastructure—particularly military assets used for communications; ISR; and precision positioning, navigation, and timing.[43] Moreover, deterrence should not only be thought of from a military perspective but also incorporate economic and diplomatic points of leverage. While integrated deterrence, and the ability to respond across domains is important, there are key advantages to being able to response in kind, tit for tat, in the same domain, and immediately, as this avoids gaps in an escalation ladder, enabling clear signaling and avoiding inadvertent escalation.

Space Domain Awareness

For the USSF to achieve any of its objectives, being able to collect information and conceptualize the threat environment is essential. The military's ability to conduct operations "detecting, tracking, and identifying all artificial objects in Earth orbit" has traditionally been known as space situational awareness (SSA), with the term referring to enabling "the continuous preparation of the battlespace in order to fight and win a war in space."[44] However, once the U.S. Air Force recognized space as a contested domain, the term morphed to *space domain awareness* (SDA).[45] When the change occurred, the definition was updated to read, "identification, characterization and understanding of any factor, passive or active, associated with the space domain that could affect space operations and thereby impact the security, safety, economy or environment of our nation."[46]

Regardless of terminology, the importance of this USSF role cannot be understated. Tracking and categorizing threats in traditional orbits around Earth and space traffic management are just the initial dataset that the USSF will be forced to operate with.[47] As mentioned previously, "[H]ostile satellites anchored at Lagrange points would be very difficult to identify and track without using SDA assets at or near these points, allowing them to stay 'parked' for long periods. Losing track of a potentially hostile satellite could be catastrophic if it is later employed against a critical space system."[48] The USSF will need to ensure there are proper assets in place to gather information near Earth, on the Moon, and in Cislunar and deep space as American adversaries continue to advance their space programs.[49] The Air Force Research Laboratory is in the process of designing

Cislunar Highway Patrol System satellites to aid in space monitoring. The deployment of satellites to address all these missions will need to be prioritized and pursued over time.[50]

USSF FINDING ITS FOOTING

The new military branch is facing growing pains while simultaneously finding areas where it flourishes. The USSF is working toward developing a distinct culture, finding ways to be supported by the National Guard and reserve forces, providing training and doctrine for Guardians, and defining ways to collaborate with other government agencies.

USSF Culture

Having a clear mission and operating procedures is important, but culture is what drives employees' thinking and behavior. While the USSF pulled military members from each service, it is largely an outgrowth from the U.S. Air Force and as a result may embrace the Air Force vision and culture.[51] One mechanism proposed to help alter the culture, for example, was for the USSF to adopt ranks utilized by the Navy due to similarities between the sea and space domains, but senior USSF leaders—who had not supported the USSF's creation—kept Air Force ranks with only a few slight deviations for enlisted Guardians.[52]

Each of the services has its own culture, which has been shaped over generations and fits with the service's mission. For example, the Air Force is known for a culture of "centralized control, decentralized execution," compared with the U.S. special operations forces that require a high degree of autonomy to act in a rapidly changing environment, resulting in a "can do" culture.[53] For the USSF to be successful, it will likely need to step outside the shadow of the Air Force and operate within a culture that matches the "can do" attitude and "encourages the focus, thinking, and behaviors required to maximize comprehensive national space power."[54] The "can do" culture will allow the USSF to be innovative and respond to threats rather than be tied to a system of capabilities and requirements like that of the Air Force.[55]

The USSF will have an opportunity to make cultural changes in other areas as well. For example, the USSF is now charged with managing the procurement of multibillion dollar satellite contracts.[56] According to Rep. Jim Cooper, chairman of the House Armed Services Subcommittee on Strategic Forces, "[T]he Space Force now has an opportunity to make real change to an acquisition culture that has been mired in cost overruns, schedule delays, and delivery of systems that are not adequately protected to survive the environment they will have to operate."[57] On a positive note, the USSF has already tried "new approaches to acquisitions, such as rapid prototyping of satellites, which allows the government to adjust requirements and incorporate the latest technologies before systems go into production."[58]

USSF National Guard and Reserve Components

The USSF remains in an awkward position because, unlike other military services, it does not have a National Guard or reserve component. Moreover, because the USSF still falls within the jurisdiction of the Air Force, the USSF leadership does not have full managerial control over personnel decisions for space professionals in the Air National Guard and Air Force Reserve. In the National Defense Authorization Act for Fiscal Year 2022, lawmakers did not approve the creation of a USSF National Guard or reserve unit.[59] The White House released a statement arguing that

establishing a Space National Guard would not deliver new capabilities—it would instead create new government bureaucracy, which the Congressional Budget Office estimates could increase costs by up to $500 million annually. The Air National Guard and Air Force Reserve units with space missions have effectively performed their roles with no adverse effect on [the Department of Defense's] space mission since the establishment of the Space Force.[60]

This focus on tactical performance exaggerates the cost and misses the strategic rationale.[61]

In the Air National Guard's current configuration, approximately 2,000 guard members perform space domain-related operations.[62] Though their numbers are small, many of these operators wield a comparably oversized amount of experience due to their full-time jobs working in private-sector space firms.[63] According to Christopher Stone, former special assistant to the deputy assistant secretary of defense for space policy in the Pentagon, the National Guard members

provide nearly three-fourths of the Space Force's overall warfighting capacity and 20 percent of its manpower. . . . It is inexplicable that the current construct has them disconnected from the active space service's organize, train and equip plans and processes, given that space guardsmen remain under the Air Force's chain of command. This invites undue risk that is wholly self-inflicted, akin to the dysfunction of placing the Air National Guard in the Department of the Army or the Army National Guard in the Department of the Navy.[64]

Furthermore, Stone argues that the disorganization impacts "training, readiness, . . . and this lack of organizational and fiscal alignment has seriously affected missions such as space electronic warfare, space surveillance and satellite operations."[65] Similarly, the plan to form a USSF reserve unit fell victim to the same fiscal scrutiny. The reserve units that perform space operations are responsible for 26 percent of the total USSF mission, but the tasks are performed by members of other military branch reserve units.[66]

The point of contention is not that National Guard and reserve units under the USSF would not be effective. At this time, the Government Accountability Office and the Office of Management and Budget have assessed them not to be cost effective, though this is at odds with National Guard Bureau estimates. The narrow focus on personnel cost ignores the unique strengths a reserve unit could bring to the USSF. A USSF National Guard or reserve unit could focus recruiting efforts on young engineers and space professionals in the commercial sector and space startups in ways that the active-duty USSF cannot. Reserve USSF Guardians can provide the vital link between the service and America's dynamic space sector. They can also play a key diplomacy role through the State Partnership Program. Policymakers cannot ignore this opportunity when considering a USSF reserve component.

Training and Doctrine

As questions regarding the surge capability support for the Guardians are addressed, training and doctrine are areas that military members operating in the space sector must focus on. Now that space is considered its own domain, the Guardians will require training that was either not necessary or not considered while the space domain was part of the air force and other service space missions. The USSF is designing a professional military education (PME) that is specific to space warfare and with space-domain-centric courses. PME programming is available through Airman Leadership School, a new USSF Noncommissioned Officer Leadership Academy, a larger Schriever Space Scholars Program at Air Command and Staff College, and the newly founded West Fellowship for Senior Developmental Education.[67] Additionally, the foundational doctrine for space operations identified five core competencies for Guardians that are in line with both current and future missions.[68] In the future, education will need to be expanded beyond the satellite mission curriculum to include supporting space commerce and exploration missions.[69] Here are the necessary U.S. Space Force spacepower core competencies:

- Space Security: Establishes and promotes stable conditions for safe and secure access to space activities for civil, commercial, intelligence community, and multinational partners.
- Combat Power Projection: Integrates defensive and offensive operations to maintain a desired level of freedom of action relative to an adversary. Combat Power Projection in concert with other competencies enhances

freedom of action by deterring aggression or compelling an adversary to change behavior.

- Space Mobility and Logistics: Enables movement and support of military equipment and personnel in the space domain, from the space domain back to Earth, and to the space domain.
- Information Mobility: Provides timely, rapid, and reliable collection and transportation of data across the range of military operations in support of tactical, operational, and strategic decision making.
- Space Domain Awareness: Encompasses the effective identification, characterization, and understanding of any factor associated with the space domain that could affect space operations and thereby impact the security, safety, economy, or environment of our nation.[70]

Advances in training and doctrine and a distinct culture for USSF personnel are good steps toward strong retention and the creation of a resilient force. However, the Guardians will also need to have adequate surge capacity through the National Guard or reserves to bolster the ranks. Furthermore, it will be important to "cultivate the next generation of space talent and ensure that they have the opportunity to pursue careers in the space sector. The USSF will need to create opportunities for its officers to advance and receive promotions in order to retain personnel."[71]

USSF Government Partnerships

For the USSF to successfully perform its mission, it must positively engage and collaborate with all U.S. government agencies either directly or tangentially related to both commercial and military space sectors. The USSF and the Defense Department more broadly will also need to foster partnerships within the private sector.

In the Defense Department and the intelligence community, several agencies operate in the space domain, so clear delineation of responsibilities is essential for smooth operation. To that end, in August 2021, several space-related agencies in the Defense Department and the intelligence community signed a classified space warfare agreement known as the Protected Defense Strategic Framework.[72] Gen. James Dickinson, commander of U.S. Space Command, characterized the agreement by stating:

Chinese and Russian space activities present serious and growing threats to U.S. national security interests. To further streamline our unity of effort, United States Space Force, United States Space Command, and National Reconnaissance Office developed a framework focused on strategic-level collaboration that increases national security collaboration. The memorandum formalizes end-to-end coordination between the Department of Defense and Intelligence Community, and between acquisitions and operations.[73]

In further collaboration across the Defense Department, the USSF is focused on providing a durable missile-warning apparatus—in combination with combatant commands, the Missile Defense Agency, the National Reconnaissance Office, and the Space Development Agency—to track incoming missile threats to the U.S. homeland and American and allied forces overseas.[74] The USSF is also involved in ensuring America has a robust survivable nuclear command, control, and communications architecture and national security space launch to guarantee that military and intelligence assets can get on orbit.[75] Strong coordination across national security enterprises is critical, but it is equally important that the same level of coordination is present across the civil sector.

Within the U.S. government, NASA is the premier civil space agency and a natural partner for the USSF to coordinate space activity and work together with foreign governments on space issues.[76] Unlike the USSF, which serves to protect and defend U.S. capabilities in space, NASA has a mission to "explore . . . the unknown in air and space, innovate . . . for the benefit of humanity, and inspire . . . the world through discovery."[77]

To better define areas of collaboration, the two entities signed a memorandum of understanding in September 2020. According to the memo, NASA and the USSF agreed to a partnership citing the following areas of common interest:

1. Deep space survey and tracking technologies to support extended [SDA] and [near-Earth object] detection beyond geosynchronous orbit;
2. Detection and data collection on bolides caused by natural objects entering Earth's atmosphere to provide timely reporting to the public and the scientific community;
3. Capabilities and practices enabling safe, sustained near-Earth and Cislunar operations such as communications; navigation; space structure servicing, assembly, and manufacturing; and interoperability among those capabilities to support resilience for functions in this remote region;
4. Search, rescue, and recovery operations for human spaceflight;
5. Launch support;
6. Space logistical supply and support;
7. Ride shares and hosted payloads to and beyond Earth orbit;
8. Establishing standards and best practices for safely operating in space, to include conjunction assessment, SSA sharing, orbital debris mitigation, and space systems protection;
9. Interoperable spacecraft communications networks for Earth orbit and beyond;
10. Fundamental scientific research and technology development cooperation; and
11. Developing and sharing a talent pool of premier space professionals and expertise.[78]

The Department of Commerce is another civil entity focused on space development that coordinates with the USSF. In a joint statement from acting secretary of the Air Force John Roth and Gen. John Raymond, USSF chief of space operations, the USSF will focus on "shared interests including space traffic management, positioning, navigation, and timing programs, applications, and efforts to maintain the space industrial base."[79] The coordination with the Commerce Department will become increasingly important as space commerce increases over the next few decades. The USSF, NASA, and the Commerce Department will also need to work with industry to form a strong public-private partnership and leverage commercial solutions.[80]

THE EVOLUTION OF THE USSF

The creation of the USSF has been a monumental and necessary step to guarantee American superiority in the space domain, but the service has a limited mandate and limited funding to match. For the USSF to become the military service necessary to maximize American spacepower, it will need to broaden its mandate extensively and expand beyond being a satellite-based service that merely supports terrestrial forces.[81]

A Broadened Mandate

Advancing American spacepower strategically envisions the USSF securing[82] and advancing America's economic interests in space,[83] providing security services for human activity in Cislunar space (including space resource extraction),[84] and acting in a manner similar to a coast guard or a blue-water navy in space. Some of the potential future roles of the USSF include planetary defense, space rescue operations, space-based air and missile defense, space traffic and debris management, space refueling, and the defense of space commerce.

Regarding enhanced SDA, the USSF will need to focus particularly on space traffic management and updating sensor systems to track spacecraft trajectories in nontraditional orbits and naturally occurring space debris that could pose a security threat to Earth or U.S. space assets—this is imperative to support a role in planetary defense.[85]

Communication is key in space. The USSF will need to expand the current Earth-focused communication architecture to provide Cislunar coverage and eventually deep space coverage through the creation of interplanetary relay satellites that can aid in space mining and future exploration missions for crewed and autonomous crafts.[86]

To properly guard against potentially hostile actors on the Moon and protect commerce in the space domain, the USSF will need to adopt a mission set similar to the Coast Guard and Navy.[87] This means that Guardians will be necessary to help protect future Lunar outposts, possibly reside on space stations as they increase in size and scope, and potentially perform personnel support and security as humans begin to build orbital hotels and settlements in the domain.[88] Furthermore, as humans are more present in space, it may be necessary to conduct search and rescue missions.[89] Although it may not be necessary in the near term, the USSF will likely need astronauts, particularly since multiple nations are planning Lunar bases in the next decade.[90]

USSF Starcruiser

For the USSF to successfully expand its mission set, vehicles will be required to place Guardians in a position to respond to threats in the space domain, transit the domain, provide military support, and serve as first responders when necessary.[91] Without a means of transportation in the space domain, it will not be possible to monitor adversary activities and maneuver and control the environment.[92] Currently, the USSF relies on the X-37B space plane to conduct uncrewed missions, but the plane's orbital maneuver capabilities are limited.[93] Reusable rockets will allow for the USSF to consider more advanced space vehicle options. Some space experts have suggested that the SpaceX Starship (or similar craft)

could be used as a mobile, versatile reconnaissance platform, using its store of fuel and six vacuum-optimized Raptor engines to maneuver . . . [and strike] at the space assets of enemy nations in times of war and defending American satellites and other space-based installations. The rocket ship could refuel American satellites, extending their operational lifespans. It could even be used to help clean up space debris.[94]

Game-changing heavy reusable space vehicle rocket technology could be instrumental in bolstering the USSF's future missions if launch costs reach $2–$5 million per launch.[95] Space strategists are now conceiving of scenarios in which the USSF builds a "starcruiser"—a navy cruiser–equivalent for space—to carry out future crewed and uncrewed missions.[96] Although a future with humans in space for prolonged periods may appear to be decades away, these technologies take significant time to develop, so the time for research and development is now.[97]

FUNDING AND THE FUTURE

American national security, and the United States' growing list of space-based economic assets, requires a committed military presence with the capability to ensure that space-based systems remain protected from dangerous naturally occurring phenomena (including asteroids and comets) and potential adversaries (like Russia and China) who are actively developing the means to disrupt, degrade, and destroy vital components of the critical and emerging U.S. space architecture. While the United States has begun moving in this direction with the December 2019 founding of the U.S. Space Force, America's newest military branch still needs a clear mission and a better-defined set of objectives. Moreover, the USSF will need funding commensurate with its mission set.

The new USSF's operating budget was $15.4 billion for fiscal year 2021 and increased to about $18 billion for fiscal year 2022. While the funding was sufficient to get the service started, the budget will certainly need to be increased to expand the USSF mission accordingly. During an American Foreign Policy Council–organized briefing for congressional staffers, Space Force Association president Col. William Woolf said,

In recent years, NASA has been able to cut costs and improve efficiency by outsourcing certain roles to the private sector (e.g. space launches). Similarly, the USSF will potentially be able to outsource certain mission sets to the private sector as well; however, it will first need to conduct a capability gap assessment to determine the service's resources and needs, and identify outsourcing opportunities.[98]

The USSF would greatly benefit from contributing to the creation of new technology that can help to pay for its funding, similar to the GPS constellation, which has an incredible return on investment.[99]

Congress has the unenviable job of deciding how much funding is necessary to maximize the USSF's effectiveness. Space strategists and the USSF leadership working in concert with the intelligence community and private sector will need to keep U.S. policymakers informed of likely threat scenarios and areas of opportunity for investment to help with funding prioritization for the force.[100]

The USSF is just one sector in a comprehensive U.S. space vision. If the United States does not develop a long-term space strategy, it risks failing to anticipate new threats and lacking the capabilities needed to advance and defend U.S. interests in space. The USSF will be an integral part of that strategy, and a broad mandate will be beneficial for USSF strategic planning and creativity. The USSF near-term priorities are not necessarily problematic, but the USSF needs to bolster its SDA and better prepare for potential conflicts in space. To ensure that the USSF develops the capabilities it envisions in the future, Congress must further codify roles and missions it will be called on to play and fund the USSF accordingly.[101]

Chapter Highlights

- **Rationale for a new military service.** Space was unique in that it was both a separate physical medium and a separate legal domain. Yet it had no service dedicated to excellence in spacecraft or the development of domain-specific expertise. To secure U.S. long-term strategic and economic interests, there must be a uniformed force—analogous to the U.S. Navy or Coast Guard on the high seas—to provide security for commerce.

- **USSF priorities.** Established near-term priorities include space domain awareness; satellite operations and global, integrated command and control of military space forces; global and theater military space operations to enable joint campaigns (to include missile warning); space support to land, air, naval, and cyber forces; spacelift and space range operations; space-based nuclear detonation detection; and prompt and sustained offensive and defensive space operations to achieve space superiority.

- **Striving for spacepower.** Accruing spacepower allows a country to better accomplish its objectives through space activity. Retaining spacepower requires having the military might to foster and protect commercial development in space and to continuously generate wealth-producing national power.

- **Space military operations.** The United States is capable of launching several types of counterspace capabilities, of varying destructive ability, on both space and ground infrastructure. Deterrence should incorporate economic and diplomatic points of leverage.

- **USSF finding its footing.** The USSF is working toward developing a distinct culture, finding ways to be supported by the National Guard and reserve forces. The USSF must positively engage and collaborate with NASA, the intelligence community, the Department of Commerce, and the private sector to be successful in its mission.

- **The evolution of the USSF.** Advancing U.S. spacepower strategically envisions the USSF securing and advancing U.S. economic interests in space, providing security services for human activity in Cislunar space (including space resource extraction), and acting in a manner similar to a coast guard or a blue-water navy in space. Potential future roles include planetary defense, space rescue operations, space-based air and missile defense, deploying space-based weaponry, space traffic and debris management, space refueling, and the defense of space commerce.

- **Funding and the future.** Congress must decide how much funding is necessary to maximize the USSF's effectiveness. Space strategists and the USSF leadership working in concert with the intelligence community and the private sector will need to keep U.S. policymakers informed of likely threat scenarios and areas of opportunity for investment to help with funding prioritization for the force.

CHAPTER 6

Shaping the Global Rules-Based Order of Space

Peter A. Garretson, Richard M. Harrison,
Lamont Colucci, and Larry M. Wortzel

Today, global norms in space, particularly on issues such as the weaponization of space, resource rights, removal of space debris, and space commerce, remain fluid and largely unformed. As the United States secures its position as a space leader, it should foster relationships with allied nations, broaden its relationships with nontraditional partners in space (India, the United Arab Emirates, Nigeria, Brazil, and the like) and increase the pace of progress toward common objectives and standards in the space domain. Broad partnerships will allow Washington to more effectively shape global norms in space.

There are multiple templates for creating space norms, one of which would use international law to develop space law. The term *international law* is misunderstood by policymakers, academics, and the general public. Much of international law, and hence future space law, is based on diplomatic custom and culture, which is rarely enforceable on belligerent "law breakers." Even those parts of international law that are governed by conventions and treaties can only be enforced by a signatory willing to use force or coercion. An excellent modern example is the international norm against invading another nation. Russia clearly broke that norm in invading Ukraine, and it was sanctioned but not forcibly stopped by other nations. International law is enforceable only if there is force behind it. Another example is the Philippines going to the International Court of Arbitration over China's occupation of Scarborough Shoal in the early 2010s. The court found in favor of the Philippines.

The Chinese government said it didn't recognize the decision and continued to occupy the shoal.

Treaties are the "hardest" form of such law, the nature of which was set out in the 1969 Vienna Convention on the Law of Treaties—they are binding and are based on the "good faith" of the participants. Diplomatic custom is somewhat softer, but it includes agreed-upon norms such as diplomatic immunity and the sovereignty of foreign embassies.

In general, conflict over future space norms will likely focus on the same sources of Earth-based international law: namely, jurisdiction and sovereignty. In general, outer space is thought of as the high seas, where claims of sovereignty, denial of access, and restrictions on use are prohibited.

One of the single greatest issues in the space norms debate will come from the core debate regarding norms on Earth. Namely, Russia and China are notorious for signing agreements, agreeing to norms in theory (regarding, for example, arms control and human rights), and then completely disregarding them in practice. Thus, it becomes increasingly difficult to have the trust necessary to sign space agreements and establish norms when there is no effective enforcement mechanism in place to deter bad behavior—aside from naming and shaming. This inability to trust U.S. adversaries such as Russia and China will likely be the bedrock problem for space governance in the future.

KEEPING THE SPACE DOMAIN SAFE AND SECURE

Despite the complexity of dealing with potentially nefarious actors, as technology continues to advance at record pace across space sectors, policy and norms will need to be instituted in the following areas, among others, to ensure a space domain that is safe and secure: territorial, property, and resource rights on asteroids and celestial bodies; the creation and removal of space debris in orbit; space traffic management, rendezvous and close approach to satellites and collision avoidance protocols; the treatment and care of the space environment; and planning for planetary defense.

Navigating the Ungoverned Expanse

Although space is an infinite domain with plenty of room for all nations to get a large slice of the expanding economic pie, arriving first is essential for setting internationally agreed-upon norms with the ability to enforce them. Allowing China to continue on a path to prominence in space is a dangerous one given the nation's propensity to ignore international law. For example, the Moon offers valuable commodities at specific locations, like water in the polar regions. According to a report by the U.S.-China Economic and Security Review Commission, "[I]n 2015 Ye Pejian, the head of China's Lunar exploration program likened the Moon and Mars to the

Senkaku Islands and Spratly Islands, respectively, and warned that not exploring them may result in the usurpation of China's space rights and interest by others," which demonstrates that Beijing is already focused on claiming Lunar resources.[1]

If Beijing and Moscow are able to team and establish a presence on the Moon before the United States, they will "begin to establish de facto norms of behavior, enabling them to 'set the tone' for future lunar operators. . . . By setting acceptable standards of behavior early, actors who subsequently join the lunar surface may be more likely to follow these norms or would at least be viewed as breaking the norms."[2] (There are reports that China expects its uncrewed Lunar station to be complete in 2027—eight years earlier than planned.)[3] The benefits of establishing norms are applicable not only to ensuring access to vital sections of the Moon but also to significant Cislunar orbit regimes like the strategic Lagrange points, as discussed in chapter 3.[4] However, territorial and resource rights in space are just one of many issues on which the international community will need to reach a consensus on in the coming decades.

Managing Space Debris

Space is an exceedingly vast domain, but there are a limited number of desirable orbits that satellites transit in around Earth. There is a misconception that because space is so large, there is no danger of satellite collisions; however, this is not the case. In late 2021, for example, China complained to the United Nations about having to maneuver its space station away from a SpaceX Starlink satellite.[5] There are over 4,000 active satellites in orbit, and millions of pieces of space debris from previous launches and destructive anti-satellite (ASAT) tests. Also, in late 2021, the State Department confirmed the Russians conducted an ASAT test that left 1,500 pieces of debris and caused the International Space Station (ISS) astronauts, some of whom are Russian, to shelter in their crew capsules.[6] In response to a November 2021 Russian ASAT test, Vice President Kamala Harris announced that the "United States commits not to conduct destructive, direct-ascent anti-satellite (ASAT) missile testing, and that the United States seeks to establish this as a new international norm for responsible behavior in space."[7] The new norm is a positive development if the United States is able to get adversarial nations to conform to it and if there is a penalty for further kinetic ASAT testing.

Likewise, as a means to combat the amount of space debris from ASAT tests, U.S. secretary of defense Lloyd Austin released a memo titled "Tenets of Responsible Behavior in Space."[8] The unclassified memo called for spacefaring nations to "1) Operate in, from, to, and through space with due regard to others and in a professional manner; 2) Limit the generation of long-lived debris; 3) Avoid the creation of harmful interference;

4) Maintain safe separation and safe trajectory; and 5) Communicate and make notifications to enhance the safety and stability of the domain." [9] Interestingly, due to the phrasing of "long-lived," the memo may actually incentivize nations to conduct ASAT weapons tests if the resulting debris from an ASAT test will deorbit quickly.[10] However, without any codification on ASAT tests, the clutter will continue to accumulate.

With so much space debris, some experts warn about the Kessler syndrome (also called the Kessler effect). The Kessler syndrome is a scenario in which low Earth orbit is so densely populated that when objects collide, the incident potentially triggers multiple collisions, creating a multitude of space debris that drastically increases the odds of future collisions.[11] The final result could be the loss of whole orbital regimes, or possibly even the inability to launch through the debris.

To manage the problem of space debris, an international consensus will need to be reached to start actively removing objects—particularly, those in common orbital lanes. One regulatory concern is "that there is no law of salvage[12] as there is at sea. At sea, if you abandon a vessel and someone finds it, it is salvage and they can keep it. In orbit, if you abandon a spacecraft and 30 years later someone approaches it from another country, it is thought of as an act of war."[13] Clarity will be necessary to determine which deactivated satellites are allowed to be removed and by whom.[14]

It is not yet clear if private-sector companies will attempt to profit from active debris removal or if countries will use their space program resources and who will pay for it. However, it is within all nations' best interest to declutter space, and proper legislation will be necessary to clean up space debris and to deter nations and private companies from creating debris in the first place.

Managing Space Traffic

As policymakers consider proposals for how to manage and inevitably reduce the amount of debris in space, it is also important to find ways to govern the satellites that are already in orbit and those that will be launched in the coming years. With ever-dwindling launch costs and a reduced barrier to entry to put a satellite in orbit, the skies are becoming crowded. The SpaceX Starlink satellite network was developed "to provide low-cost internet to remote locations. SpaceX eventually hopes to have as many as 42,000 satellites in this so-called megaconstellation."[15] If this projection is realized, SpaceX would own 10 times the number of active satellites in orbit today! The sheer number of satellites will undoubtedly increase the likelihood of potential collisions (called conjunctions). In addition, such a large number of satellites passing across the sky will cause problems for astronomers.[16]

The increased volume of satellites will necessitate collision avoidance protocols. Who is responsible for determining when two satellites are on a collision course? And if a collision is projected to occur, which satellite is expected to alter course? Satellites have a finite amount of propellant aboard, so every orbital adjustment shortens the vehicle's life. Space experts have suggested reviewing how commercial and military ships interact at sea as a template for maneuver decisions in space.[17] If one of the satellites is no longer active, for example, satellite owners will need to be clear on whether they can move the dead satellite instead of altering course.[18] Other experts have proposed the use of transponders, as in airplanes, and developing auto-collision-avoidance software for satellites.[19] The space situational work related to collisions is currently governed by the U.S. military, but this may be a role adopted by the Department of Commerce in the future.[20] In addition to collision avoidance, there are no rules or regulations governing the ability to fly within a certain distance of a satellite, known as close approach.

Shaping the Space Environment

As leaders across the globe consider how best to mitigate climate change and the negative impacts humans have had on Earth, it is important to consider these lessons as humans move out into space. The space environment will need some rules of the road to ensure that, when companies or countries are extracting resources from the Moon or launching from the Lunar surface, these actions are coordinated with others to mitigate any potentially long-lasting environmental impact to the Moon or other celestial bodies. The California Gold Rush offers a perfect analogy for unregulated expansion into a new frontier:

Lawless competition proceeded as actors exploited new resources without accountability. In California, the abundant resources deteriorated as the newcomers dammed rivers, chopped trees, and mined the land. Gold Rush anarchy served as an impetus for building mining infrastructure and for signing a new state constitution. With governance providing surety, commercial actors further exploited minerals for economic gain. The United States tamed the chaos.[21]

Likewise, the United States has an opportunity to lead in shaping space environmental norms.

Planning for Planetary Defense

Much focus has been placed on the debris created by launches and ASAT tests; however, it is at least as important to consider threats from naturally occurring space artifacts, such as asteroids. These large space rocks can hurtle toward Earth and cause human extinction level events. The need for planetary defense is a topic that both China and the United States are beginning to take seriously.

A body of literature in China recognizes that asteroids once did enough damage to Earth to wipe out a significant amount of life and create numerous ice ages. In one case, research from 2019 speculates that a major asteroid strike in what is now South Carolina and other places in the United States approximately 13,000 years ago caused "major disruption and extinction," wiping out woolly mammoths, among other species.[22] A conference on space in Guangzhou, China, focused primarily on extracting resources from space bodies like planets and asteroids, also noted the potential danger to Earth posed by major asteroid strikes that might be averted by intercepting and redirecting the asteroids.[23]

A *Xinhua* article from 2020 tracked NASA's research into its Double Asteroid Redirection Test (DART)—a technology demonstrator designed to explore dynamic impact technology and permit a spacecraft to "intercept near-Earth asteroids, thereby changing the asteroid's orbit, and ultimately preventing it from hitting the Earth."[24] China's January 2022 white paper on space specifically mentions "study[ing] plans for building a near-earth object defense system, and increase[ing] the capacity of near-earth object monitoring, cataloguing, early warning, and response."[25]

The United States is already past the planning stages for planetary defense. NASA's Planetary Defense Coordination Office has chosen to conduct a kinetic impactor demonstration by hitting Didymos, a binary near-Earth asteroid, to change its course.[26] This is an area where the U.S. Space Force may play a larger role in the future.[27] Moreover, commercial operations may be an option for planetary defense support as space miners could deflect an asteroid as part of a mining operation.[28] Nevertheless, the threat from asteroids is not unique to any one country, and this may be an area in which to consider international cooperation to protect the planet.

SPACE TREATIES AND AGREEMENTS

International law records agreements between nations and defines (albeit loosely) formal norms of behavior that states are expected to respect. Inside many nations, including the United States, a formal treaty once ratified is treated as domestic law. However, between states, the situation is different. Unlike domestic law, there is no government above nations with enforcement power. The power of international law comes chiefly from the interest most states have in maintaining a reputation among other states as being reliable and trustworthy in their word and agreements and being a member in good standing in the international community. As is seen in other areas of international law (Russia in Ukraine, China in the South China Sea), nations can ignore international law. However, the utility of international law is to encourage prosperous economic and social cooperation and to provide clarity regarding who breaks community norms (which impacts the offending party through reputational costs).

International space law, as it is currently known, can best be described as four treaties dating from the Cold War under the U.N. umbrella and an attempt to demilitarize space and create a broad international regulatory framework that was not ratified by any major world power. The four major U.N. conventions are managed by the U.N. Office for Outer Space Affairs.

In summary, the U.N. treaties are the Outer Space Treaty of 1967, preventing placement of weapons of mass destruction (WMD) in space and forbidding national claims on space territories; the Agreement on the Rescue of Astronauts, the Return of Astronauts and the Return of Objects Launched into Outer Space of 1968, an agreement defining parties an obligation to render assistance; the Convention on International Liability for Damage Caused by Space Objects of 1972; and the Convention on Registration of Objects Launched into Outer Space of 1976, governing registration of space objects, which holds nations responsible for their space objects (see table 6.1 for more details). The Agreement Governing the Activities of States on the Moon and Other Celestial Bodies, the so-called Moon Agreement of 1979, remains unratified by any major spacefaring nation, like the United States or China. The Moon Treaty would effectively ban resource harvesting, colonization, and claims on the Moon. The Prevention of an Arms Race in Outer Space resolution of 1981 is also unratified and is opposed by the United States—this proposal attempts to ban all weapons in space, not just WMD. More recently, the Artemis Accords, which were first signed by the United States and several partner nations on October 13, 2020, appears to be a push by the United States and allies to set its interpretation of U.N. space treaties.

Outer Space Treaty

The Outer Space Treaty was signed in 1967 and created to serve the interests of the two major powers at the time: the United States and the Soviet Union.[29] Its primary utility was in cementing an existing understanding that it was in neither superpower's interest to proliferate nuclear weapons to outer space, which would involve a costly arms race, lower reaction times, and introduce uncertainties into command control. It enabled both superpowers to limit the expense of competing on the Moon and to mitigate the corresponding fear that if the other party got their first, they might claim it all. Furthermore, it also reflected the need for both superpowers to get the support of newly independent former colonies.

Table 6.1 International Space Agreements, 1967–Present[i]

U.N. Outer Space Treaty 1967[ii]	• The Treaty on Principles Governing the Activities of States in the Exploration and Use of Outer Space, including the Moon and Other Celestial Bodies, or U.N. General Assembly Resolution 2222 (XXI), entered into force in October 1967. The treaty states that the use of space shall be peaceful, that states shall not deploy WMD into space, that states shall not make claims of sovereignty in space, and that states are responsible for damage they cause in space. • The United States, Russia, and China are parties to the treaty.
U.N. Rescue Agreement 1968[iii]	• The Agreement on the Rescue of Astronauts, the Return of Astronauts and the Return of Objects Launched into Outer Space, or U.N. General Assembly Resolution 2345 (XXII), entered into force in December 1968. • The agreement states that astronauts who land in another state's territory are to be rescued and provided aid as necessary in the event of an emergency. Space objects that land outside of the owning state's territory are also to be returned by the receiving state. Compensation is allowed for rescue missions.
U.N. Liability Convention 1972[iv]	• The Convention on International Liability for Damage Caused by Space Objects, or U.N. General Assembly Resolution 2777 (XXVI), entered into force in September 1972. The convention provides for states to make claims against other states for damages caused by their space launches or space objects and a commission for settling said complaints. • The convention holds that states are responsible for damages caused by their space objects. The United States, Russia, and China are parties to the convention.
U.N. Registration Convention 1976[v]	• The Convention on Registration of Objects Launched into Outer Space, or U.N. General Assembly Resolution 3225 (XXIX), entered into force in September 1976. The convention holds that states must register any space object launch with the United Nations and provide key launch and orbital details and must inform the U.N. when space objects deorbit. • The U.N. Office of Outer Space Affairs manages the space object registry. The United States, Russia, and China are party to the convention, but China does not register satellites with the United Nations.

Table 6.1 (*continued*)

U.N. Moon Agreement 1984[vi]	• The Agreement Governing the Activities of States on the Moon and Other Celestial Bodies or UNGA Resolution 34/68 entered into force in July 1984. The agreement reiterates the key points of the U.N. Outer Space Treaty but focused on the Moon.
	• The agreement states that the Moon shall be used only for peaceful purposes and that no WMD or military bases will be placed on the Moon or in orbit around the Moon. It attempts to establish an international authority to control mining activity. The United States, Russia, and China are not parties to the agreement.

Note: WMD, weapons of mass destruction.

[i] The authors would like to thank AFPC junior fellow Cody Retherford for generating the table information.

[ii] "Treaty on Principles Governing the Activities of States in the Exploration and Use of Outer Space, including the Moon and Other Celestial Bodies," U.N. Office of Outer Space Affairs, Resolution 2222 (XXI), 1966, https://www.unoosa.org/oosa/en/ourwork/spacelaw/treaties/introouterspacetreaty.html.

[iii] "Agreement on the Rescue of Astronauts, the Return of Astronauts and the Return of Objects Launched Into Outer Space," U.N. Office of Outer Space Affairs, Resolution 2345 (XXII), 1967, https://www.unoosa.org/oosa/en/ourwork/spacelaw/treaties/introrescueagreement.html.

[iv] "Convention on International Liability for Damage Caused by Space Objects," U.N. Office of Outer Space Affairs, Resolution 2777 (XXVI), 1971, https://www.unoosa.org/oosa/en/ourwork/spacelaw/treaties/introliability-convention.html.

[v] "Convention on Registration of Objects Launched Into Outer Space," U.N. Office of Outer Space Affairs, Resolution 3235 (XXIX), 1974, https://www.unoosa.org/oosa/en/ourwork/spacelaw/treaties/introregistration-convention.html.

[vi] "Agreement Governing the Activities of States on the Moon and Other Celestial Bodies," U.N. Office of Outer Space Affairs, Resolution 34/68, 1979, https://www.unoosa.org/oosa/en/ourwork/spacelaw/treaties/intromoon-agreement.html.

While the treaty's primary utility has been successful in creating a global norm against the proliferation of nuclear weapons into space, its downside has been to dissuade human expansion and commercial space development by removing one of the most powerful incentives—the ability of states to acquire territory that enables private individuals to own and develop real property.

The Outer Space Treaty is technically ambiguous on the topic of nonnuclear weapons in space. It does not prohibit space-to-space weapons, ground-to-space weapons, or even space-to-ground weapons unless they are WMD. To be clear, the treaty prohibits the placement of WMD in Earth orbit. However, the treaty does not prohibit military personnel on the Moon for peaceful purposes. It does prohibit placing armaments used to conduct military maneuvers on the Moon, but it provides no mechanism for resolving disputes, only a "request for consultations."

The Outer Space Treaty is also best understood in the context of an arms control treaty, not as a treaty that is designed to enable commercial space development and multinational cooperation for space governance.[30]

Artemis Accords

More recently, the 2020 Artemis Accords are a serious cornerstone for space norms. The accords are an initiative by the United States to provide clarity regarding how to execute the Outer Space Treaty to enable not only exploration of the Moon but also its use. It is an attempt to build a community-wide system of norms for how to operate as part of a coalition of nations exploring and developing the Moon. While the specific accords are bilateral agreements between states cooperating on the Artemis program, the Artemis principles[31] are open for anyone to subscribe to, and it is hoped that they will become a global norm, followed even by adversarial states not taking part in the multinational program. Importantly, the Artemis Accords codify the U.S. interpretation of the legality of the use of space resources. At the time of writing, 20 nations and the Isle of Man have signed on to the Artemis principles, and all are participating in the Artemis program. The power and caliber of signatory states is significant. In aggregate, they account for more than 45 percent of the global GDP.[32] The accords have been signed by senior officials for space policy for Australia, Bahrain, Brazil, Canada, Colombia, France, Isle of Man, Israel, Italy, Japan, South Korea, Luxembourg, Mexico, New Zealand, Poland, Romania, Singapore, Ukraine, the United Arab Emirates, the United Kingdom, and the United States.[33]

The Artemis Accords represent a multinational agreement centered on the Moon, but they have implications for other celestial bodies. They include an agreement that extraction and utilization of space resources should be conducted in a manner that complies with the Outer Space Treaty and in support of safe and sustainable activities. The signatories affirm that this does not inherently constitute national appropriation, which the Outer Space Treaty prohibits. They also express intent to contribute to multilateral efforts to further develop international practices and rules on this subject.

The accords are widely viewed as an attempt by the United States and its allies to define principles related to mining of the Moon and other bodies, but the benefits of these efforts can be enjoyed by all nations. Adversaries, by contrast, are attempting to create norms for their own benefit.

Treaty on the Prevention of the Placement of Weapons in Outer Space and the Threat or Use of Force Against Space Objects

China and Russia have a long history of cooperation in space, even if that cooperation was interrupted during the 1960s and 1970s.[34] Beginning in the 1950s, cooperation with Russia, and then the Soviet Union, was important for the development of China's space program.[35] China received missiles and technical help from Russia, and Russia allowed China to put scientists into its missile program. Both forms of assistance aided in the development of China's DF-1 ballistic missile, which was first launched November 1960.[36] However, the Sino-Soviet split in August 1960 led to the end of that cooperation. Sino-Russian cooperation began again in the 1990s after the collapse of the Soviet Union, although it was a marriage of convenience, as it allowed the Russian space industry to remain buoyant and provided China much-needed access to Russian technology to further its progress in space.

More recently, the two countries have contested norms and coordinated their positions on space in the United Nations. According to a space law expert, "[T]he most significant attempt at creating an arms control agreement for outer space is the Sino-Russian 2008 Draft Treaty on the Prevention of Placement of Weapons in Outer Space and the Threat or Use of Force Against Space Objects" (PPWT).[37] Further, "[t]his proposal of a legally binding multilateral agreement on the issue of preventing an arms race in space, submitted to the Conference on Disarmament, failed to generate adequate support, both in 2008 and when a new version was submitted in January 2014."

The U.S. position on the PPWT is that the draft treaty is flawed. It has no verification mechanism, it does not restrict the development of ground-based ASAT weapons, it does not address directed energy weapons such as lasers that can destroy or disable a satellite temporarily, and it does not ban the launch of direct-ascent ASAT weapons.[38] Still, the PPWT is an excellent example of the geostrategic policy cooperation between China and Russia on space.

The clearest example of cooperation between the two countries today is the announcement that China and Russia will partner in the founding of a permanent Moon base.[39] While neither a treaty nor an agreement, China and Russia's partnership offers an alternative to shaping exploration and use of the Moon through their proposal for an International Lunar Research Station (ILRS). They have provided a road map and user's guide for their endeavor:

CNSA and ROSCOSMOS jointly invite all international partners to cooperate and contribute more to the exploration and use of the Moon in the interests of all humankind adhering to the principles of equality, openness, and integrity. CNSA and ROSCOSMOS provide a series of cooperative opportunities for all interested international partners in the phases of the plan, design, development, implementation, operation and scientific research of the ILRS project.[40]

There are several concerning strategic implications of the proposed road map, including that Russian and China would gain "access to launch sites, ground stations, and receiver stations in China and Russia, as well as access to a universal scientific talent pool, to include growing Chinese and Russian space expertise, and burgeoning employment opportunities in China where aerospace salaries are becoming globally competitive. They will also be able to divide the long-term costs of research and development."[41] Moreover, the two nations are strengthening their space ground game on Earth:

As of May 2022, the Prague Security Studies Institute (PSSI) has identified globally 303 Chinese and Russian transactions targeting 83 countries. 14 of those transactions are between China and Russia. We also identified 12 "international" transactions (which we define as international multilateral agreements sponsored by Russia and/or China). Out of the total number of recorded (bilateral) transactions China accounted for 147 transactions targeting 71 countries, Russia for 130 transactions targeting 43 countries.[42]

The United States will need to rapidly develop a holistic plan that counters China and Russia on both Earth and the Moon to effectively compete in strategic space initiatives.

Key International Institutions Governing Space Policy

The key international bodies related to outer space governance include the U.N. Office for Outer Space Affairs (UNOOSA), the U.N. Committee on the Peaceful Uses of Outer Space (UNCOPUOS), and the Conference on Disarmament. UNOOSA is a U.N. office staffed by U.N. officials. UNCOPOUS is the main forum for nations to interact and includes a group of governmental experts. The Conference on Disarmament is the key forum where previous arms control agreements have been negotiated. The International Space Exploration Coordination Group is an assembly of 27 participating agencies that attempt to coordinate and advance a global exploration strategy.

Important nongovernmental institutions that have helped advance space development include space advocacy groups such as the Alliance for Space Development, the National Space Society, the Space Frontier Foundation, and international groups such as the Hague International Space Resources Governance Working Group and the International Institute of Space Law.

Of interest, there are no equivalent development organizations analogous to how the United States shaped the post–World War II world—and there should be. There is no space equivalent of the International Civil Aviation Organization to reconcile and publish best practices and aid in traffic control. There is no equivalent to the World Bank or the International Monetary Fund to provide financing for space development. There is no equivalent to the Organization for Economic Cooperation and Development to coordinate space development policies. Further, there is no equivalent to the North Atlantic Treaty Organization to provide collective defense in space.

The majority of space policy has been governed by international bodies with nations as the voting members and legislators. However, today, activity in space is dominated by the private sector, which owns the majority of satellites (only 1 in 10 satellites are government owned)[43] and is projected to have an overwhelming number of satellites in the coming half decade (approximately 100,000 by 2030).[44] Space experts have noted that as international institutions govern space, the private sector will need to have a seat at the table or, at a minimum, representation from their host countries to ensure they are participating in the discussions (unfortunately, the Outer Space Treaty regime makes states responsible and liable for all commercial activities rather than assigning responsibility and liability directly).[45]

OPPORTUNITIES FOR U.S. COLLABORATION IN SPACE

While the United States contemplates expanding its ambitions in space, there is no reason for it to do so alone. The head of U.S. Space Command, Gen. John Raymond, has made it clear that working with like-minded countries is now a necessity: "Historically, we haven't needed to have allies in space. . . . [Allied partnership] is a big growth area for us. And I think it's going to provide our country a big advantage. We're stronger together."[46] This view should not be unique to the military. NASA, in particular, presents a strong vehicle for collaboration, given that it has a history of teaming with other countries on space initiatives.[47] One of the comparative advantages the United States enjoys is a strong alliance portfolio relative to China and Russia.[48] In an era of great power competition, the United States should be further nurturing these alliances as a strategic asset. This naturally should include space, as partners and allies are also releasing their own space strategies and have their own concerns about the increasingly contested nature of space, which is likewise crucial to their economies, society, and military.[49] Moreover, allied partnerships hold increased importance because potential adversaries may collude on space initiatives; indeed, Russia and China have opened discussions on future space cooperation.[50]

Department of Defense Allied Collaboration

NASA is always viewed as the U.S. government's best asset for space coordination with foreign governments; however, the Department of Defense should not be overlooked. The Defense Department organizes numerous training and partnership programs worldwide with other militaries, and, with so many nations forming space forces,[51] the U.S. Space Force has a great opportunity to partner in similar fashion. At the same time, building architecture together and having mutually supporting capabilities is another avenue that should be pursued.

Former vice chairman of the Joint Chiefs Gen. John E. Hyten said that American defense satellites were "big, fat, juicy targets," which highlighted the need to move toward a more nimble and resilient space architecture.[52] The military's ability to diversify and distribute information and space-based capabilities across multiple U.S. and allied satellites to complicate adversary targeting choice is critical.[53] Also, more narrowly, it can serve broader deterrence purposes if the United States collocates military assets on allied satellites, as it raises the risk of horizontal escalation if Russia or China were to strike a satellite with collocated assets.[54] Moreover, recent war games suggest that allied cooperation is key to resolving space conflicts.[55]

These types of initiatives are already occurring. "For example, a hosted payload agreement with the government of Norway will save [the United States] more than $900 million and helped us get capability on orbit two years faster."[56] This type of cooperation serves to reduce cost and increase the pace of innovation. Another example "is the development and integration of space domain awareness payloads on two Japanese Quasi-Zenith Satellite System spacecraft. The hosted payloads will increase sensor diversity and enable space surveillance and event detection over USINDOPACOM in the geosynchronous orbit regime."[57]

The Artemis Gateway project is an example of a planned collaborative endeavor. The project, which will serve as a Lunar outpost, will have contributions from the Canadian Space Agency, the Japanese Aerospace Exploration Agency, and the European Space Agency.[58] The development is a significant international statement of unity and a bulwark to outstrip the Chinese and Russian planned activities.

In addition to international collaboration, the United States will need to consider interpretations of current treaties and the impacts on space. For example, if the United States and China are in a space war, does that trigger NATO's Article 5? In other words, does collective defense apply?[59] NATO has already declared that space is an operational domain and issued a statement saying that attacking a satellite irreversibly is considered an act of war.[60] Further cooperation between the Defense Department and other countries is welcome, but broader U.S. government cooperation with allies will be necessary for primacy in space.

Bilateral Space Cooperation

The United States has an opportunity to shape space as a domain of free and open commerce. That shaping requires a multilateral agenda to create multiple new institutions that (a) coordinate, fund, and regulate space economic and infrastructure development; (b) create institutions that establish a collective defense community specific to space, including planetary defense; (c) lead and coordinate specific projects such as Lunar development and space solar power; and (d) build consensus and establish norms and governance that moderate potential conflict and clarify objectionable interactions between states in the space domain.

Bilateral opportunities for developing space include the opportunity to use space to build long-term relations, integrate markets, and offer U.S. services (commercial, civil, and national security) to meet development needs. To realize these opportunities with developing states, the United States requires (a) a strategy to offer linked GPS, information technology, and financial and infrastructure systems to counter China's Belt and Road Initiative space information corridor for development commensurate with U.N. Millennium Development Goals; (b) a strategy to build basic space competence in state planning and basic capacity to operate and build satellites; (c) a more coordinated "playbook" and FAQ for engagement; and (d) deliberate consideration of space in country plans. Certain developing states may also offer geographic access for uplink/downlink stations, space domain awareness sensors, and future spaceport locations. Fortunately for the United States, there is no shortage of countries and regions that are potential parties for it to engage with in space policy discussions and development.

India Opportunities

India is the major long-term cooperation opportunity for U.S. space policy and the one relationship that will provide the greatest opportunity to shape the future. India is the largest democracy with the greatest potential for economic growth in the long term—potentially even eclipsing China and the United States by 2060.[61] India is a mature spacefaring power, with independent capability to manufacture satellites and launch them to low Earth orbit and geosynchronous orbit. It has successfully sent probes to the Moon and Mars and is working on both reusable launch vehicles and a human spaceflight program. India is also a nuclear-capable power and an ASAT-capable power with its own dedicated military space organization. India and the United States have shared concerns about China's ambitions and military actions. The recent signing of a U.S.-India agreement for space domain awareness and exchange of military space personnel is a major step forward,[62] but much more is possible.[63]

Japan Opportunities

Japan is among America's closest allies with the most advanced space capabilities, including the ability to independently build spacecraft and launch them. Japan has created a dedicated military space unit,[64] is a formal treaty ally, participates in U.S. space war games, and shares U.S. concerns about China's territorial ambitions in the near abroad and in space.

Moreover, the nation has signed on to the Artemis Accords and will likely be an important participant in Lunar development. Recently, Japan passed a space mining law, signaling interest in that sector. Additionally, Japan has long had ambitions to advance space solar power[65] and has recently committed to a demonstration by 2025.[66]

Europe Opportunities

Europe is home to many of America's closest allies who share a vested interest in space as well as concerns about security and climate. European partners have already signaled their interest in being part of America's Lunar plans. NATO has declared space an operational domain,[67] and France is developing counterspace capabilities and a dedicated military space unit.[68] Countries such as Luxembourg are pushing the boundaries of space finance and property rights,[69] supporting the U.S. interpretation of the use of space resources.

Middle East Opportunities

Across the Middle East, numerous countries are now beginning to build space programs with grand ambitions. The United Arab Emirates is the first country in the Middle East to send a Martian probe and the first nation to articulate an ambition to colonize Mars with a city.[70] Additionally, the UAE has also been a thought leader on space finance, space resources, and property rights.

Africa Opportunities

Africa represents an untapped zone of opportunity for American space cooperation.[71] Several African countries now have space agencies, and multilateral efforts have led to the creation of a pan-African space agency.[72] In the coming decades, many African nations will see significant development, and, as a result of Africa's size and infrastructure challenges, space services can play a unique role in African development. Furthermore, space information services such as precision navigation and timing and space internet are intimately linked with the broader societal information and communications systems and infrastructure. And with more than a billion people in a growing population, Africa is being targeted by China for investment in a space information corridor to lock African nations into China's standards and suppliers. The United States will need a more complete whole-of-government strategy to compete.[73]

South America Opportunities

The United States has an interest in the welfare and development of South America. With a common heritage of having won freedom from colonial powers and melting together diverse peoples into new nations, we share an interest in a connected hemisphere. In South America, several nations are developing basic spacefaring capabilities, and all nations need space information services for development. Brazil, in particular, has a blossoming space program and a large developing economy poised for takeoff.[74] The United States can extend its existing commercial and security relationships to ensure that it is the preferred partner for space-enabled development and in-space development across the South American continent.

DEFINING ROLES WITH COMPETING NATIONS IN SPACE

For many years after the collapse of the Soviet Union, the United States and Russia achieved a successful collaboration on the ISS. Russia was a capable, and largely reliable, partner on the ISS. However, even prior to the Ukraine invasion, Russia demurred on joining the Artemis program only to join China as the junior partner in China's ILRS. Further, Russia tested a debris-causing ASAT weapon that endangered the ISS and its own astronauts. Under President Vladimir Putin, Russia appears to be pulling away from the transatlantic community and its norms. Russia's invasion of Ukraine and the sanctions it has generated will complicate any future space cooperation—Russia has already threatened to end support for the ISS because of sanctions related to the Ukraine conflict.[75]

Relations between China and the United States on space exploration and space launch were not always competitive. From the Chinese perspective, U.S.-China space cooperation was seen as a way to stabilize bilateral relations in the 1990s.[76] However, since Congress passed the Wolf Amendment in 2011, NASA has been specifically forbidden from spending U.S. funding to cooperate with China on space initiatives.[77] Although some observers in NASA and the United States think there should be areas of cooperation in space, there are strong arguments that such programs compromise U.S. security and technology.[78]

While Americans are generally idealistic and hopeful about the possibility of international cooperation, even with competitors, several issues complicate collaboration with China. Many idealists start from the hope that if the United States chooses a symbolic overture of space cooperation, the signaling of a friendly intention on the ground stage of space will lessen tensions and reduce the problems inherent in the security dilemma. Their hope is that enmeshing the two countries' space programs can stabilize the relationship, create friendships, and raise the costs for defection. They point to historical examples such as the Apollo-Soyuz mission and the ISS, which provided highly visible symbols of cooperation.

However, the ISS also demonstrates the downsides of such cooperation when there are conflicts on Earth (as with Russia's invasion of Ukraine), as well as the difficulty of maintaining such cooperation or even access if one party is dependent on the other. The recent examples between the United States and Russia with regard to the ISS should give one pause. Cooperation in space does not produce cooperation on Earth but merely reflects the degree of Earthly alignment, and that can evaporate quickly. There are other considerations that should give one pause when considering whether to cooperate and how to cooperate with China.

Even if the United States' primary interest is to avoid a space arms race because it fears it might lose the race, fears the cost, fears it might lead to war, or fears it might lead to accidental escalation, then American leadership must remain clear-eyed about the effects of cooperation on other U.S. interests and weigh the relative costs and benefits. A Sino-U.S. space partnership would be concerning for several reasons.

First, any symbolically important cooperation on a major aspirational goal would essentially put the United States and China on equal footing. This will undermine the perception of American leadership and standing in the world, and China would certainly use it to say, "We have arrived. Even the former hegemon recognizes it." America, in fact, would have to kowtow and acknowledge the power of China. Any symbolically important cooperation legitimizes the Chinese Communist Party and its actions more broadly, and the United States loses the moral high ground. America would in effect put bad behavior at the front of the line and put other countries that are attempting to mimic American values behind them, signaling that what matters to America is power.

Second, cooperation would have consequences for global alignment. Once the United States itself is cooperating with China on space, it opens the door and essentially gives permission to all U.S. allies and friends to cooperate with China. That weakens a U.S. coalition and further traps U.S. allies into dependency on China.

Third, by entrenching the United States in a codependent relationship on a highly visible space project, the United States puts its own political capital at risk any time it criticizes China on human rights or its behavior around the world because China might defect from the common program. The United States would be placing its own goals at the mercy of a power with deeply different interests and values. And China is notorious for expressing its displeasure with diplomatic slights by constricting access to key dependencies.

Fourth, U.S. policymakers must contemplate how such cooperation affects overall American competitiveness. Space is the source of many innovations that proliferate to other fields. Countries often engage in shared projects to avoid incurring the total cost for the project so that they can put their energy someplace else. If the United States were to relinquish 50 percent of an ambitious space program to China, that means 50 percent less activity for U.S. industry and a reduced share of potential innovation at home. Moreover, the People's Republic of China has a long history of state and corporate espionage to steal intellectual property from the United States, and this has been a major deterrent to cooperation. In fact, past commercial operations have been used to materially advance China's nuclear weapons strike capability and counterspace capability that now target the United States.[79] Close cooperation means enhanced access and exposure to the United States' best practices, insights, and intellectual property and increased opportunity for espionage.

In sum, symbolic cooperation is likely to reduce the status of the United States, weaken an American-led coalition, hold the U.S. space program hostage to the Chinese government's sensitivities, and undermine the American industrial and military advantage. The net effect of cooperation is likely to advance China's relative position at the United States' expense with very little hope of reducing fundamental adversarial postures, competition, and conflict.

Although symbolic cooperation may not be possible, nor in U.S. interests, there are areas where the United States and China may have shared interests in avoiding unwanted tensions. For example, all nations will benefit if there are fewer weapons causing space debris. Therefore, it may be in the interests of both nations to close the club of ASAT-missile-capable nations with a control regime. Likewise, both nations might agree that use of an electromagnetic-pulse (EMP) weapon or a high-altitude nuclear detonation (HAND) device could be deemed illegal or a war crime.

There are also a few opportunities to engage on issues involving private-sector and civilian considerations. For example, it may be of mutual interest to disambiguate or restrain competitive desires of private industry to drag nations into conflict through a registry of claims. Additionally, it may be possible to cooperate on some level of space traffic management for commercial satellite traffic, allowing a separate system for military spacecraft that gives due regard to commercial traffic. Finally, consideration should be given to developing standard practices and standards (emergency frequencies, docking standards, airlocks) that allow for mutual aid and assistance in emergencies.

Chapter Highlights

- **Keeping the space domain safe and secure.** Sound policy and norms are needed to ensure a space domain that is safe and secure in relation to, among other things, territorial, property, and resource rights on asteroids and celestial bodies; the creation and removal of space debris in orbit; space traffic management, rendezvous and close approach to satellites and collision avoidance protocols; treatment and care of the space environment; and planning for planetary defense. An international consensus is needed to start actively removing space debris, an increased volume of satellites will require collision avoidance protocols, it is important to consider long-lasting environmental impacts of space mining on the Moon and celestial objects, and planetary defense is a potential area of international cooperation.

- **Space treaties and agreements.** The U.N. treaties are the Outer Space Treaty of 1967, preventing placement of WMDs in space and forbidding national claims on space territories; the Agreement on the Rescue of Astronauts, the Return of Astronauts and the Return of Objects Launched into Outer Space of 1968, an agreement that defines parties' obligation to render assistance; the Convention on International Liability for Damage Caused by Space Objects of 1972; and the Convention on Registration of Objects Launched into Outer Space of 1976, governing registration of space objects and holding nations responsible for their space objects. The Artemis Accords are an initiative by the United States to provide clarity regarding how to execute the Outer Space Treaty on the Moon not only to enable exploration but also to codify the legality of using space resources. The private sector owns most space assets, and it needs to help govern space policy.

- **Opportunities for U.S. collaboration in space.** Bilateral opportunities for developing space include the opportunity to use space to build long-term relations, integrate markets, and offer U.S. services (commercial, civil, and national security) to meet development needs. As the United States expands its ambitions in space, there is no reason for it to do so alone. One of the comparative advantages the United States enjoys is a strong alliance portfolio, relative to China and Russia.

- **Defining roles with competing nations in space.** A Sino-U.S. space partnership would be concerning for several reasons. Symbolic cooperation is likely to reduce the status of the United States, weaken an American-led coalition, hold the U.S. space program hostage to the Chinese government's sensitivities, and undermine the American industrial and military advantage. However, both nations could ban space EMPs, nuclear detonations, and ASAT weapons and agree on space traffic management, standard practices for docking, and emergency assistance.

CHAPTER 7

Charting the Dimensions of Space Competition

Peter A. Garretson and Richard M. Harrison

The new space economy has the potential to grow to trillions of dollars annually in the coming decades. In the unfolding era of great power competition, the United States will be dealing with finite resources, making it optimal to develop a plan that maximizes a return on investment. For example, there may not be much strategic value in going to Mars in the near term, unless it is to claim territory for future human settlement. By contrast, mining the Moon may be a worthwhile endeavor in the nearer term, because extractive industries ("space mining") may have both strategic and commercial returns. The Chinese government has already outlined a strategic plan to develop comprehensive national power via space.[1] Moreover, as of December 2021, China has accelerated its plan.[2] For America to adequately compete in the space domain, it will be important to understand which sectors of the space domain offer the most strategic and economic opportunities.

When determining how best to prioritize various opportunities to pursue in space, it is important to consider views from experts across various space sectors. New Space New Mexico convened a virtual State of the Space Industrial Base 2020 conference held in partnership with the U.S. Space Force (USSF), the Defense Innovation Unit, and the Air Force Research Laboratory, making a strong contribution to space policy.[3] The conference brought together 120 individuals from private-sector companies, the Pentagon, and other government agencies and departments (including NASA, the Department of Commerce, and the Department of Energy) to discuss economic and military leadership in space. The conference produced a number of key recommendations designed to provide a foundation for America to become the world's predominant space power.[4] In particular, the conference report concluded the following:

[There are] six areas vital to overall U.S. national spacepower and the U.S. space industrial base and the areas most likely to be centers of gravity in great power competition:

- **Space policy and finance tools** to secure U.S. space leadership now and into the future by building a unity of effort and incentivizing the space industrial base.
- **Space information services** include space communications/internet, precision positioning, navigation, and timing (PNT), and the full range of Earth-observing functions, which have commercial, civil, and military applications.
- **Space transportation and logistics** to, in, and from Cislunar space and beyond.
- **Human presence** in space for exploration, space tourism, space manufacturing, and resource extraction.
- **Power for space systems** to enable the full range of emerging space applications.
- **Space manufacturing and resource extraction** for terrestrial and in-space markets.[5]

The report covers the state of each sector and the associated challenges, offering general recommendations for the short, near, and long terms. As such, the report lays a sound foundation for space policy. Yet, several points of elaboration will prove beneficial for policymakers, including (a) how to prioritize pursuing these recommendations across the six space sectors, especially given the finite limits to available resources; (b) understanding the degree to which the United States benefits economically or strategically from specific efforts within each space sector; and (c) understanding how the United States compares to other nations, notably China, in each sector. In this chapter, the six centers of gravity are evaluated to better understand which are the most promising to give the United States a competitive advantage over its adversaries.

To provide clarity on strategic and economic priorities in space, the American Foreign Policy Council's Space Policy Initiative engaged dozens of space experts from across academia, government, and the private sector via a survey and recorded interviews. Their insights provided useful guidance for a comparative assessment of the United States and China with respect to each critical space sector.

In *Developing National Power in Space*, space strategist Brent Ziarnick asks a pressing question: "How do we know whether a space development project truly advances strategic access and a space power's ability?"[6]

Ziarnick's answer: "[A] project advances strategic access if it will open a legitimate market, area, or natural resource to exploitation by commerce to generate wealth on a permanent cost-effective basis."[7] This provides a useful framework to consider in terms of prioritizing the various space sectors.

EVALUATION OF SIX CENTERS OF SPACEPOWER COMPETITION

Space Policy and Finance Tools

The enabling or disabling posture taken by states via policy and finance tools determines the availability of resources and the friction or ease associated with public and private entrepreneurs' efforts to develop space. When effective space policy and finance measures are in place, states can pursue innovations to build a space economy and national spacepower, along with the transportation systems, power, extractive industries, and information systems that propel them.

China has, by far, the most coherent and comprehensive set of policies for the economic development of space and a goal of achieving space primacy by 2045.[8] Their policies start with a vision of becoming the preeminent space power and using the vast resources of space to build a $10 trillion Moon-Earth economic zone by 2050.[9] This vision is driving specific investments in enabling technologies such as space solar power,[10] space nuclear power and propulsion, in-space extractive industries,[11] in-space manufacturing, and closed-cycle life support and space agriculture to enable settlement. The Chinese State Council Information Office released a white paper, titled "China's Space Program: A 2021 Perspective," which officially discusses "the steps it will start a new journey towards a space power."[12]

In addition to substantive policy planning, the People's Republic of China (PRC) is proverbially "putting its money where its mouth is" by investing in space. From 2019 to 2020 alone, China's investment in its aerospace sector more than tripled, increasing from $296 million to $933 million.[13] The PRC is able to harness private-sector support in coordination with its government programs due to its military-civil fusion (军民融合) strategy, which blends its "defense industrial base and its civilian technology and industrial base."[14] The military strategy includes exploiting the U.S. dependence on space through robust counterspace programs.[15]

In contrast, the United States has taken some proactive measures but is far from unleashing the full suite of tools available to fully compete on the final frontier. As detailed in several reports,[16] the United States needs a North Star vision and execution plan to fully unlock the resources required to build a Cislunar economy. Although America has yet to articulate a comparatively grand strategy for space, Congress is funding and NASA is executing the Artemis program to push for Lunar exploration and the

Commercial LEO Destinations project to guarantee continuity after the International Space Station (ISS) sunsets—NASA has already awarded $416 million for this private space station endeavor.[17] Moreover, several enabling policies and plans have been set forth by the Defense Innovation Unit, the Commercial Satellite Communications Office (CSCO)/Commercial Services Office,[18] and budding research and development budgets across the military to develop the key components of an in-space industry. These efforts are designed to spur innovation by rapidly commercializing new ideas that affect defense and security in space, among other domains.

On the economic front, there are a few proposed initiatives to rapidly stimulate space growth. As the USSF evolves and engages in longer-duration missions, in-space refueling may become a requirement that promotes prospects for long-term space development. Congress has long asked the Department of Defense to develop a responsive launch capability. Now that there are many options for commercial launch, the USSF could purchase rides to space as a "launch on schedule" at a regular cadence. This would provide regular and predictable rides to space to enable a rapid tempo of innovation. Should payloads not be ready, the USSF could launch propellant, enabling on-orbit refueling, extending the life of current satellites by years.

Experts have recommended that the USSF have a mandatory percentage of commercial buys and a $1 billion initial fund[19]—with the project executed through expansion of the USSF commercial communications purchase. The proposed U.S. Space Commodities Exchange should help foster further development of the Cislunar economy by providing propellant resources for purchase.[20]

To further propel the U.S. space economy, it will also be imperative to get projects off the ground with adequate funding. Positively, the private sector has seen a large influx of capital as space ventures attracted $15 billion in financing during 2021—over double the year prior.[21] However, much of this investment is focused on near-term applications, especially for large constellations of smallsats (miniaturized satellites) in low Earth orbit (LEO), with substantially less investment toward infrastructure to enable the next increment of the space economy, including energy, manufacture, and in-space logistics. Some space experts have suggested creating a federally charted space corporation that "will be self-sustaining and have the authority to issue bonds and offer optional finance mechanisms to develop space into the future" to help create and incentivize companies to join the space economy.[22] Others have proposed developing a strategic materials reserve in space that incentivizes companies to mine asteroids.[23]

In the aggregate, the American space projects are supportive of advancing U.S. spacepower but lack the coordination and focus necessary to stay ahead of China. As explained by spacepower theorist Joshua Carlson in

Spacepower Ascendant, nations can explore, expand, exploit, and potentially exclude.[24] While the United States is busy exploiting LEO for short-term economic gains, China is focused on expansion into Cislunar to set itself up for long-term economic success through efforts like its International Lunar Research (ILRS) Station to create in-space production capacity. Carlson likens this to the British efforts to set up economically productive and self-sustaining colonies and towns in North America, which was ultimately more successful than Spanish pillaging of New World gold.[25] Thus, in the absence of a broader vision to help coordinate efforts, insufficient attention is being paid to long-term investments.

Space Information Services

Space information services—which includes PNT; overhead sensing (optical, infrared, multispectral, radar, and radio-frequency mapping); navigation and timing signals; and communications (satellite TV, satellite radio, satellite data, and broadband internet)—make up a large swath of the global space economy. The decrease in launch costs continues to enable growth in this space sector.

China is a fast follower on space information services, rushing to catch up with the United States. The PRC has rapidly populated its own national satellite communications system, is competing against GPS with its BeiDou Navigation Satellite System,[26] rapidly populated an Earth-observation system,[27] and is now seeking to compete for global space-based broadband internet.[28] For example, the BeiDou system is a strategic and economic asset, as the system is used in 120 countries, which deepens these nations' dependency on Beijing, and BeiDou is projected to produce over $150 billion in value by 2025.[29] China offers it as part of the Belt and Road Initiative (BRI), a strategic offering that helps cement the power bloc the PRC is forming. The Chinese government's State-Owned Assets Supervision and Administration Commission announced a satellite broadband internet project that is equally ambitious, as it envisions 13,000 satellites providing global internet coverage that will likely be used to further entrench other nations' reliance on China.[30]

Despite a well-earned reputation for cybertheft,[31] the PRC has shown a capacity for innovation. It built the world's first space-based quantum-key encryption system;[32] was the first to extend a communications system to the far side of the Moon,[33] enabling access to the shadowed regions of the Lunar poles; and, most recently, added its Lunar satellite relay to the Very Long Baseline Array, now the world's largest radio telescope.[34] China is encouraging its private sector[35] to supplement national and state-owned companies to build new space information services. The fact that China is now innovating should be of great concern to the United States—there was a time when the United States copied many innovations from Europe, but eventually the United States became known as the center of global innovation. If the United States were ahead in all space activities, China would be copying us, but in fact, it has surpassed us on some fronts.

While China continues to rise, the United States maintains the world's most advanced space information services. Today, GPS navigation systems are the gold standard and are relied on globally. Studies show that since its inception in the 1980s, GPS has generated well over $1.4 trillion in economic value, and a potential GPS service outage could have a negative impact of approximately $1 billion per day.[36] In addition to its economic importance, the system is strategically vital, as the U.S. military relies on it heavily for navigation, positioning of troops, and targeting, among other applications—and the system is continuing to modernize and increase its robustness.[37] Moreover, the vitality of the U.S. space private sector is sky high, with massive networks of commercial information-gathering and disseminating constellations. For example, SpaceX is able to launch more than 60 Starlink satellites at a time, and Planet Labs has more than 200 satellites orbiting Earth and collecting and distributing Earth-imagery data.[38]

In terms of a commercialized, commoditized service, communications are the most mature sector, providing a model for the development of other services. A mature system exists for government, especially for military purchases of commercial satellite communications services.[39] Overall, the United States remains well positioned to operate in the space information services sector.

Space Transportation and Logistics

The ability to move easily, inexpensively, and rapidly into space and around Cislunar space will be an invaluable asset for any nation.

China has orchestrated a plan to develop its space transportation and logistics system. It aspires to convert its entire rocket fleet, including for heavy launch, to full reusability before 2035.[40] The PRC also appears to be attempting an almost direct copy of the SpaceX Starship.[41] If a similar spacecraft is realized before SpaceX, the Chinese will gain a strategic advantage and compete for global point-to-point transportation opportunities. Moreover, for both economic and strategic purposes, the PRC has plans to develop nuclear shuttles to give it access to the asteroid belt,[42] and Beijing is actively encouraging its private sector to construct launch vehicles of all varieties—particularly, reusables. The Chinese are utilizing their newly launched Tiangong space station to perfect crewed rendezvous and docking operations and to enhance their ability to navigate in near-Earth orbit.[43] Farther out in space, China's plans for the Moon assume industrial scaling using extractive industries and additive manufacturing (3-D printing).[44]

Meanwhile, in the United States, the private sector has taken the lead in innovations in space transportation and logistics (STL), also called space access mobility and logistics (SAML). Both NASA and the Defense Department have played an enabling role, helping to develop fundamental technologies and provide contract opportunities to build logistics capabilities with wider market applications. Positive civil examples include NASA Commercial Orbital Transportation Services (COTS),[45] NASA Commercial Crew,[46] NASA Commercial Lunar Payload Services (CLPS),[47] NASA's In-Space Manufacturing (ISM) project,[48] various NASA innovative advanced concepts,[49] and NASA centennial challenges.[50] Positive Defense Department examples include the national security space launch (NSSL); USSF's SpaceWERX Space Prime;[51] the Defense Innovation Unit's commercial launch, multi-orbit logistics,[52] and outpost;[53] and the Defense Advanced Research Projects Agency's (DARPA) DRACO program[54] and its novel orbital and Moon manufacturing, materials, and mass-efficient design (NOM4D) program.[55]

Despite these positive steps, NASA has not created a more encompassing "Lunar COTS" program that catalyzes broader logistical efforts and architecture for a sustainable and affordable Cislunar transportation system. It has provided no clear vision for its sustained operations after 2028 and provided no set plans for when and how it will convert from its expensive and expendable systems to reusable commercial systems. Moreover, NASA has not outlined plans involving partners to build and operate a "Lunar industrial facility" with contracted extractive production targets nor specified public-private partnerships to develop operational refueling capabilities, landing pads and precision landing systems, power, or agriculture.

Delays are also a concern. Artemis program timelines continue to slip—from 2024[56] to 2025[57] to 2026[58]—delaying access and momentum; undermining U.S. and allied self-confidence; and showing the United States in poor light compared to China, which meets its deadlines. It is a substantial problem that NASA has been unable to meet political timelines (and perhaps is unconcerned about its failure to do so).[59]

On the military front, neither the U.S. Space Command nor the USSF has outlined a clear architectural vision or published formal requirements to drive innovation in SAML or to use commercial purchases to advance this area.

Both civil and military projects could more deliberately drive U.S. advantage in STL and SAML. Creating enabling policies might include mandating purchase of transportation as a service versus purchasing the vehicles themselves, mandating a certain percentage to "buy commercial," providing a stable forecast (analogous to the Defense Logistics Agency fuel purchases) for logistics commodities, and the creation of a space commodities exchange. Such policies would send a much stronger demand signal to industry. The United States may also wish to look to means beyond NASA and the Defense Department to create STL, including proposals such as the Foundation for the Future's Space Corporation Act,[60] the use of "public capitalization notes" suggested by economist Armen Papazian,[61] the use of space bonds to fund infrastructure,[62] and an authority to create an in-space strategic mineral and propellant reserve,[63] all of which could greatly accelerate progress.

Human Presence in Space

Human presence in space has been symbolic, but many observers believe that achieving a sustained settlement beyond Earth is essential to still broader goals, such as making humanity and life multiplanetary and ensuring the perpetual existence of humankind. Those nations that have led in human presence have historically assumed the banner of leadership of all humanity.[64]

China has articulated a concrete objective of moving past the milestone of just visiting space to achieving human settlement in space. Toward that end, China is experimenting with growing plants on its Lunar lander and space station and is experimenting with closed-cycle life support on Earth.[65] China has plans for both Lunar and Mars crewed landings.[66] As a stepping stone, China has an advanced human spaceflight program and has built its own space station. The Chinese astronauts (taikonauts) continue to grow their human space activity acumen by increasing durations of over 90 days of crewed time in space.[67]

The PRC is also encouraging companies to build vehicles for space tourism. Beijing plans to capitalize on space tourism as it works toward capturing approximately 25 percent of the projected $1.7 billion global space tourism market by 2027.[68]

Although China is making steady progress, the United States continues to lead in human presence in space. To date, no other nation has exceeded U.S. successes in human space exploration. However, such achievements are now able to be copied. The vanguard of vision and aspiration for human presence has passed from NASA's tame desires for a sustained "campground" presence on the Moon and a human visit to Mars to Elon Musk's much more ambitious vision of humans becoming a multiplanetary civilization by building a self-sustaining city on Mars[69] and Jeff Bezos's vision of enabling a trillion human beings to live in free-flying planned communities throughout the solar system.[70] The timetable for humans to set foot on Mars has slipped. Musk initially projected a landing date of as soon as 2024 but has pushed that back to 2029 (a timeline far in advance of NASA's)—60 years since Americans landed on the Moon.[71] On the way are near-term plans for private space stations, space tourism, around-the-Moon cruises, and high-tempo suborbital and orbital human-rated spacecraft.

A key deficit in U.S. policy is that our national purpose and the purposes of the civil and military arms do not reflect this larger purpose—space settlement is not codified in law or policy.[72] As a result, it does not drive bureaucracies to make the key investments needed to enable the settlement vision: artificial gravity, closed-cycle life support, space agriculture, large habitat design and construction, and large-scale passenger transportation.

Power for Space Systems

Nothing can be moved or transformed in space without power. More power increases the ability to transform raw materials into finished goods and services, and more power allows greater mobility from the source of production to markets. Military power also depends on energy sources. From crops that provided the calories fueling the soldiers, mules, and horses of antiquity to the oil and nuclear energy that power our modern aircraft and aircraft carriers, supplies of energy have determined national advantage. So too in space. Becoming a space power without in-space power is not possible. Power today in space is anemic compared to that on Earth. The largest single platform, the ISS, generates only 120 kilowatts of power, hardly enough to build an industrial and manufacturing supply chain in space.[73] To get to megawatts or even gigawatts of power requires investment in space nuclear power and industrial-scale space solar power.

A nation with large amounts of in-space power for industrial applications also has available to it large amounts of energy to convert to military purposes if threatened. Energy powers transportation, logistics, sensing, and long-distance communication. And, yes, energy can power lasers, particle beams, high-power microwaves, and rail guns.

China is clear about its ambitions to develop in-space power of both nuclear and space solar varieties.[74] Its Moon-base plans clearly articulate a nuclear reactor at the center,[75] and the PRC has a distinctly articulated target for when it intends to have nuclear shuttles. China has been working on space solar power (SSP) for over a decade, with a clear road map for in-space testing,[76] prototyping, and commercialization. The Chinese have tied SSP to their Moon program, to Lunar industrialization, and to their $10 trillion Moon-Earth economic zone.[77] They see the Moon as a supplier of energy for sustainable development.[78] China even has confidence in its fusion program and therefore has organized Lunar prospecting to look for helium-3, a clean fusion fuel.[79] In the same way in which China mobilized its provinces to take over terrestrial solar panel production, China is mobilizing its provinces to pursue SSP—it has already broken ground in Chongqing for a Bishan solar energy receiving station.[80]

Unlike China, the United States does not have a national-level program to advance in-space power. Its efforts are fragmented across NASA, the

Department of Defense (DOD), the Department of Energy (DOE), and industry. No executive order or national strategy ties efforts together. Only modest efforts have been made in the area of space nuclear power and propulsion (SNPP). NASA and DOE developed the kilopower reactor,[81] only to abandon it in favor of other directions for Lunar surface power late in this decade.[82] DARPA is advancing a space nuclear reactor for propulsion,[83] and Congress has included a $70 million line item in the USSF budget for nuclear propulsion for Cislunar,[84] but neither DOD nor DOE has a dedicated program for space nuclear power, and DOD has articulated no requirement. Tragically, the United States has no national SSP program. Its efforts at DOE ended in the 1980s, and its efforts at NASA ended around 2006. At the request of Congress, DOD has a small Space Solar Power Incremental Demonstrations and Research program and a $66 million line item for fiscal year (FY) 2022 (and $45 million in the proposed FY 2023 budget), but this is not enough to scale to a full prototype.[85] Without policies in place advocating for nuclear power and SSP, the United States is at severe economic and military disadvantage if China is successful in its endeavors in this sector.

Space Manufacturing and Resource Extraction

A nation's strength is determined by its industrial scale of production. In World War II, the Allies were able to triumph over the Axis powers principally because the United States and the Soviet Union had a greater depth of industrial capacity. After the war, the United States remained the global economic leader and shaped the world order primarily because of its scale of production. Space offers unparalleled opportunities to expand the production frontier with billionfold greater mineral resources and billionfold greater energy resources than on Earth.[86] Those nations able to gain advantage in off-world extractive and manufacturing industries will corner new markets and gain market share, contributing to societal wealth and soft power. The efficiency of in-space production also garners a huge advantage because there is a large cost to building on Earth and bringing material to, and transporting it around, space due to the need to overcome Earth's gravity. Furthermore, nations pursuing ISM and extraction will also receive a deep war chest of capital, an industrial might that can be mobilized, a logistics and supply chain with strategic depth, and an ability to outproduce and outsupply an enemy in war.

China intends to develop both in-space extractive industries and ISM at scale. Chinese Lunar plans assume industrialization to build solar power satellites, 10,000 metric tons each,[87] utilizing Lunar mining and additive manufacturing technologies.[88] China has also considered how asteroid resources could be used to build solar power satellites.[89] Additionally, past Lunar missions have prospected for resources, including helium-3,[90] and planned Lunar missions will showcase operational testing of additive manufacturing technologies that have already been demonstrated on Earth.[91]

Furthermore, China has planned multiple asteroid missions (including the capture and return to Earth of a small asteroid for study) and space mining tests.[92] The PRC has a lab dedicated to in-space resource utilization[93] and is building a research laboratory for deep space exploration in Luxembourg[94] and its new Tiandu deep space exploration lab.[95] China's plans for in-space power and STL are tied to its vision of space industrialization at scale.

Conversely, the United States has expressed only tepid interest in space industrialization, in-space extractive industries, or ISM. Both NASA and DOD have small-dollar programs to advance ISM: NASA's ISM project[96] and DARPA's NOM4D.[97] As of 2022, no national strategy or executive order guided these efforts, and they do not approach even 1 percent of NASA's or DARPA's budget. Interest from the White House and NASA encouraging private facilities in LEO and in-space servicing, assembly, and manufacturing (ISAM) may result in new programs and investments, hopefully including free-flying or attached private space stations. A major step forward was taken in April 2022, when the Biden administration released its ISAM strategy.[98] This document seeks to promote and ignite U.S. competence in this field. Unfortunately, the strategy does not mention SSP, does not extend to extractive industry, and does not establish benchmarks and timelines, and it does not appear that the Office of Management and Budget has set aside funding to execute the strategy in the president's proposed FY 2023 budget. An important improvement would be the inclusion of an ISAM line item in the budgets for NASA, the USSF, and the National Reconnaissance Office (NRO) to allow transparency and tracking. A necessary next step is to publish a companion document for extractive industries or in-situ resource utilization.

Moreover, at present, there is neither a coherent vision nor a significant budget for LEO industrialization (NASA requested $102.6 million in FY 2022).[99] The situation is even less clear for the Moon and asteroids. While NASA was directed to put its base at the resource-rich Lunar South Pole and to build a sustainable presence starting in 2028, no vision of a Lunar industrial facility has been released,[100] and no COTS-like incentive program to build Space Act partnerships has been announced.

The first round of asteroid mining companies failed, in part, due to a lack of government incentives and support. Today, there are several startups and established companies (e.g., TransAstra, OffWorld, Lunar Resources, and Lockheed Martin) desiring to develop extractive industries and manufacturing on the Moon and asteroids. But while the United States has successfully focused on securing a legal framework for commercial in-space resource extraction,[101] and has set a key precedent,[102] no serious program exists to incentivize U.S. companies to develop in-space extractive industries or ISM at scale. Unlike in Japan and Luxembourg, it is not even clear which agency, if any, has the authority to license and supervise such an activity.

It is impossible to imagine the United States as a power without its industrial might or to imagine its global military power without its steel, shipyards, oil, uranium, and aircraft factories. A failure to develop such analogues in space means no future arsenal of democracy in a future conflict.

SPACE SECTOR ASSESSMENT

Space provides a wealth of opportunity that can be realized only if the correct policies, financing, and incentives are structured to maximize a return on space investment. The preceding overview of the United States and China within each space sector provides the status and current trajectory of each nation as it pursues its space ambitions. However, for America to remain the predominant space power, it must chart a course toward maximizing spacepower.

To give legislators a starting point to identify space sectors where the United States can and should compete (and other comparatively less important areas in which the United States could conceivably cede ground), here we provide an assessment of the most advantageous areas for the United States to pursue. This assessment is based on the aggregated wisdom from 25 subject matter experts,[103] whom we asked to rank various technological focus areas for strategic and economic impact. The results of our survey are displayed in table 7.1.

As policymakers contemplate how to move forward with space policy, table 7.1 can be used to help clarify and prioritize items that are both strategically and economically important to the country. To recap, based on the survey results, the highest overall scoring investments were as follows:

1. Space-based solar power for terrestrial markets
2. Space-based real-time tracking of moving objects such as aircrafts, ships, and cars
3. In-space refueling
4. In-space construction of large platforms in geosynchronous Earth orbit
5. Space-based high-resolution video
6. Nuclear thermal propulsion
7. Lunar mining for propellant
8. Nuclear electric propulsion
9. Lunar mining for metals
10. Asteroid mining for metals

Table 7.1 Space Activities Ranked by U.S. Strategic and Economic Value

Space Activity	Strategic Value (Avg.)	Economic Value (Avg.)	Combined Value (Avg.)	Strategic Value	Strategic Rank	Economic Value	Economic Rank	Combined Rank
Space-based solar power for terrestrial markets	2.54	2.67	2.60	Medium-high	4	Medium-high	1	1
Space-based real-time tracking of moving objects (aircraft, ships, cars)	2.68	2.39	2.54	Medium-high	1	Medium-high	2	2
In-space refueling	2.60	2.17	2.39	Medium-high	2	Medium	7	3
In-space construction of large platforms (larger than today's GEO satellites)	2.48	2.28	2.38	Medium-high	7	Medium-high	3	4
Space-based high-resolution video	2.52	2.22	2.37	Medium-high	5	Medium	5	5
Nuclear thermal propulsion	2.60	2.13	2.37	Medium-high	2	Medium	8	6
Lunar mining of propellant	2.50	2.21	2.35	Medium-high	6	Medium	6	7
Nuclear electric propulsion	2.40	1.96	2.18	Medium-high	8	Medium	12	8
Lunar mining of metals	2.09	2.08	2.09	Medium	11	Medium	9	9

(continued)

Table 7.1 (continued)

Space Activity	Strategic Value (Avg.)	Economic Value (Avg.)	Combined Value (Avg.)	Strategic Value	Strategic Rank	Economic Value	Economic Rank	Combined Rank
Asteroid mining of metals	1.87	2.24	2.05	Medium	17	Medium	4	10
Tracking of near-Earth objects with 48-hour warning for impact threats as small as 15 m	2.36	1.74	2.05	Medium-high	9	Low-medium	14	11
City-scale free-flying space settlements	1.96	1.96	1.96	Medium	14	Medium	10	12
Asteroid mining of propellant	1.92	1.96	1.94	Medium	15	Medium	10	13
Lunar elevator	1.92	1.92	1.92	Medium	15	Medium	13	14
Private/commercial crewed Lunar facility	2.00	1.67	1.83	Medium	13	Low-medium	15	15
Crewed Mars facility	2.24	1.35	1.79	Medium	10	Low-medium	18	16
Mars settlement	2.08	1.41	1.74	Medium	12	Low-medium	17	17
Private crewed space facility	1.67	1.61	1.64	Low-medium	18	Low-medium	16	18

Note: GEO, geosynchronous orbit.

It is striking how underinvested the United States is in the highest-payoff areas. Of the total $50 billion budgeted between NASA and the USSF, probably less than $1 billion (or 2 percent) is allocated to these areas. This inefficiency in expenditure results from the lack of a clear American space vision.

It is also worth viewing the space activities from a macro perspective. Table 7.2 displays the data with items arranged by space sector. When binning each technology focus area into the areas of strategic competition, we find the following: For space information services, the highest-payoff investments are space-based real-time tracking and space-based high-resolution video. For STL, in-space refueling has the highest potential payoff (though, notably, nuclear electric and nuclear thermal propulsion also directly support space transportation). All power systems scored well, including SSP, nuclear thermal, and nuclear electric. Within space manufacturing and resource extraction, in-space construction and Lunar propellant scored the highest, with strong showings for Lunar mining of metals as well as asteroid mining for metals and propellant. Notably, no human-presence areas scored highly for strategic or commercial payoff, though crewed facilities on the Moon and Mars scored the best in the humans-in-space category. Of note, of the top 10 areas (see table 7.1), 4 were in manufacturing and extraction, 3 were in power, 2 were in space information services, and 1 was in STL.

BENEFITS OF GOVERNMENT SUPPORT

One major difference between China and the United States is clarity of vision for space. With a grand plan and the benefits of a military-civil fusion strategy, China has outlined more concrete steps to achieving spacepower. Table 7.3 explores subject matter expert judgments regarding the positive role that the government could play in accelerating these technological focus areas for the United States. What is striking is how many of the advanced applications and commercial markets were judged to be realizable within a decade. Orbital space tourism, in-space refueling, private crewed space facilities, space-based tracking of moving targets, space-based high-resolution video, Lunar mining of propellant, 48-hour warning for small asteroid threats, in-space construction of large platforms, Lunar mining for metals, and private/commercial Lunar facilities *were all possible within a decade.* Nuclear electric propulsion, nuclear thermal propulsion, space-based solar power for terrestrial markets, asteroid mining of propellant, and asteroid mining of metals were all judged as possible within 15 years. Only crewed Mars facility, Mars settlement, Lunar elevator, and city-scale free-flying settlements were judged farther out, and even these *were seen as realizable within 20 years,* given proper government incentives.

Equally important is the opportunity of policy to accelerate each of these endeavors. In every case, government incentives were seen as positive,

Table 7.2   Rankings of Space Activities by Space Sector

Space Sector	Strategic Importance	Economic Importance
Space Policy and Finance Tools		
	-	-
Space Information Services		
Space-based real-time tracking of moving objects (aircraft, ships, cars)	Medium-high	Medium-high
Space-based high-resolution video	Medium-high	Medium
Tracking of near-Earth objects with 48-hour warning for impact threats as small as 15 m	Medium-high	Low-medium
Space Transportation and Logistics		
In-space refueling	Medium-high	Medium
Lunar elevator	Medium	Medium
Human Presence		
City-scale free-flying space settlements	Medium	Medium
Private/commercial crewed Lunar facility	Medium	Low-medium
Crewed Mars facility	Medium	Low-medium
Mars settlement	Medium	Low-medium
Private crewed space facility	Low-medium	Low-medium
Power for Space Systems		
Space-based solar power for terrestrial markets	Medium-high	Medium-high
Nuclear thermal propulsion	Medium-high	Medium
Nuclear electric propulsion	Medium-high	Medium
Space Manufacturing and Resource Extraction		
In-space construction of large platforms (larger than today's GEO satellites)	Medium-high	Medium-high
Lunar mining of propellant	Medium-high	Medium
Lunar mining of metals	Medium	Medium
Asteroid mining of metals	Medium	Medium
Asteroid mining of propellant	Medium	Medium

Note: GEO, geosynchronous orbit.

Table 7.3 Estimated Time to Market with and without Government Assistance

Space Activity	Average Years to Accomplish Prototype (No USG Investment/ Incentives)	Average Years to Accomplish Commercial Mission (No USG Investment/ Incentives)	Average Years to Accomplish Commercial Mission (with USG Investment/ Incentives)	Years Saved by U.S. Government Assistance (for Commercial Mission)
Orbital space tourism	4.1	5.4	4.5	0.9
In-space refueling	5.2	6.7	5.9	0.7
Private crewed space facility	5.6	7.8	6.3	1.5
Space-based real-time tracking of moving objects (aircraft, ships, cars)	7.3	10.2	7.1	3.0
Space-based high-resolution video	6.8	9.4	7.1	2.3
Lunar mining of propellant	10.5	12.0	8.8	3.2
Tracking of near-Earth objects with 48-hour warning for impact threats as small as 15 m	12.0	13.0	8.8	4.2
In-space construction of large platforms (larger than today's GEO satellites)	12.0	16.2	9.3	6.9
Lunar mining of metals	11.6	13.7	10.0	3.7
Private/commercial crewed Lunar facility	13.0	14.6	10.6	4.0
Nuclear electric propulsion	12.3	16.0	11.5	4.5
Nuclear thermal propulsion	12.5	15.7	11.6	4.1

(continued)

Table 7.3 (*continued*)

Space Activity	Average Years to Accomplish Prototype (No USG Investment/Incentives)	Average Years to Accomplish Commercial Mission (No USG Investment/Incentives)	Average Years to Accomplish Commercial Mission (with USG Investment/Incentives)	Years Saved by U.S. Government Assistance (for Commercial Mission)
Space-based solar power for terrestrial markets	14.2	17.4	12.1	5.3
Asteroid mining of propellant	17.4	18.8	13.6	5.3
Asteroid mining of metals	17.3	19.0	14.1	4.9
Crewed Mars facility	18.6	20.3	16.1	4.2
Mars settlement	21.8	22.3	18.4	3.9
Lunar elevator	22.5	22.2	18.4	3.7
City-scale free-flying space settlements	23.4	23.4	19.3	4.1
			Average	3.7

Note: GEO, geosynchronous orbit; USG, U.S. Government.

Table 7.4 Space Activity Times to Market for China and the United States

Space Activity	China (Date Achieved Based on Current Space Road Map)	USA (Date Achieved with No USG Support)	USA (Date Achieved with USG Support)
Space-based solar power for terrestrial markets	2049	2038	2033
Nuclear thermal propulsion	2040	2037	2033
In-space construction of large platforms (larger than today's GEO satellites)	2035	2037	2030
Lunar mining of propellant	2040	2033	2030
Asteroid mining of propellant	2040	2040	2035
In-space refueling	—	2028	2027

Note: GEO, geosynchronous orbit; USG, U.S. Government.

with the acceleration in time to market ranging from one to seven years. In general, a proactive stance by the government to provide the proper investment and incentives is expected to *reduce the average delay until a truly commercial product is offered by an average of 3.7 years.* For some of the truly high-payoff focus areas, such as SSP and asteroid mining for metals, the acceleration in time to market was even more significant, at more than five years.

This analysis suggests that the United States has an opportunity to realize a vast ecosystem of new markets, greatly shortening the time to market offering by as much as 5 years with the proper incentives, unleashing the most promising applications within the next decade and all of the new markets within just 20 years. The key investments are in solar and nuclear power, ISM and mining (with a focus on the Moon), in-space refueling, and two high-payoff space information services: space-based moving-object tracking and space-based high-resolution video. Importantly, table 7.4 strongly suggests that right now, with proper government incentives, the United States could beat China's announced timelines in each of these key areas to maintain its position as the premier space power—however, the United States will not achieve the desired timescale if it delays. (For more information on China's projected space accomplishments, see "China's 'Space Road Map'" in chapter 2.)

CHARTING A PATH FORWARD

The U.S. will most likely lose space superiority to China within the next decade.[104]

—Nicholas Eftimiades, senior official with the
Office of the Director of National Intelligence,
National Intelligence Council

After decades working to change how the government treats space, right now I am focused on the private sector. My venture capital firm SpaceFund has raised and invested over $20M in 19 companies—most of them frontier-oriented startups. With our newest fund, in the next year both of these numbers will double. However, it is clear to me that for New Space to succeed in opening space to America and the world **we need govspace to get it right—or both will fail.**[105]

—Rick Tumlinson, founder of SpaceFund

It is high time that the United States develop a comprehensive strategy that serves its economic, societal, and military interests. To be successful, private-sector space companies, NASA, the Department of Commerce, the Department of Defense, Congress, and the White House, along with other federal agencies, will need to work together to unlock the limitless potential in space. The analysis contained herein is intended to provide policymakers with such a North Star vision, explaining the relevant strategic context that makes space so vital, defining the strategic imperatives that should drive national policy, and outlining a clear policy consensus that will withstand multiple administrations.

Chapter Highlights

- **Evaluation of six centers of space power competition.** The PRC outlined a strategic plan to develop comprehensive national power via space. For America to compete in the space domain, it needs to understand which space sectors offer the most strategic and economic opportunities.

- **Space policy and finance tools.** When effective space policy and finance measures are in place, states can pursue innovations to build a space economy and national spacepower, along with the transportation systems, power, extractive industries, and information systems that propel them.

- **Space information services.** PNT; overhead sensing (optical, infrared, multispectral, radar, and radio-frequency mapping); and communications (satellite TV, satellite radio, satellite data, and broadband internet) make up a large swath of the global space economy.

- **Space transportation and logistics.** The ability to move inexpensively and rapidly into space and around Cislunar space will be an invaluable asset.

- **Human presence in space.** Human presence in space is symbolic, but many observers believe achieving a sustained settlement beyond Earth is essential for making life multiplanetary and ensuring the perpetual existence of humankind.

- **Power for space systems.** Nothing can be moved or transformed in space without power. Power increases the ability to transform raw materials into finished goods and services and allows for greater mobility from the source of production to markets. Military power also depends on energy sources.

- **Space manufacturing and resource extraction.** The strength of a nation is determined by its industrial scale of production. Space offers unparalleled opportunities to expand the production frontier with billionfold greater mineral resources and energy resources than on Earth.

- **Space sector assessment.** Key investments are in solar and nuclear power, in-space manufacturing and mining (with a focus on the Moon), in-space refueling, and two high-payoff space information services: space-based moving-object tracking and space-based high-resolution video.

- **Benefits of government support.** A proactive stance by the government to provide the proper investment and incentives is expected to reduce the average delay until a truly commercial product is offered by an average of 3.7 years. For some of the truly high-payoff focus areas, such as space solar power and asteroid mining for metals, the acceleration in time to first market offering is even more significant, at over five years.

CHAPTER 8

Defining an American Space Agenda

Peter A. Garretson and Richard M. Harrison

America today requires a serious strategy to compete with China in the space domain. The goal is to ensure that the United States remains the global guarantor of economic prosperity, security, and freedom. The military and security threats emanating from adversary counterspace activity are comparatively clear. Equally significant, however, is the potential threat to U.S. primacy in the global geo-economic order that could be posed by the exploitation of space by foreign nations to gain vital positional and economic advantage.

The vast economic potential of space resources represents either an opportunity to extend U.S. primacy or an opening for the People's Republic of China to speed ahead and assume the mantle of global hegemon. The U.S. Air Force's Future of Space 2060 and Implications for U.S. Strategy workshop concluded the following:

- The United States must recognize that in the world of 2060, space will be a significant engine of national political, economic, and military power for whichever nations or nation best recognize(s) the potential of space and organize(s) and operate(s) to exploit and maximize that potential.
- The United States faces growing competition from allies, rivals, and adversaries to remaining the leading nation in the exploration and exploitation of space as an expanded domain for human endeavor.
- China is executing a long-term civil, commercial, and military strategy for exploration and economic development of the Cislunar domain, to include the settlement of the Moon, with the explicit aim of displacing

the United States as the leading space power. Other nations are developing similar national strategies.

- A failure to remain the leading space power will place U.S. national power at risk. The United States and its allies must promote and optimize the combined civil, military, and commercial exploitation of space that best serves the nation's interests.
- The U.S. military must define and execute its role in promoting, exploiting, and defending the expanded commercial, civil, and military activities and human presence in space driven by industry, NASA, and other nation-states.[1]

Given China's grand space ambitions, and the potential for astronautical growth, in 2019, the U.S.-China Economic and Security Review Commission recommended that Congress direct the president to initiate a long-term economic space resource policy strategy, including an assessment of the viability of extraction of space-based precious minerals, onsite exploitation of space-based natural resources, and space-based solar power. As envisioned, this effort would include a comparative assessment of China's programs related to these issues.[2] Additionally, the commission recommended publishing a formal assessment of U.S. strategic interests in or relating to Cislunar space.[3]

The decision to better understand China's strategic intentions is vital to our nation, but—more importantly—America needs a unified front on space policy. For the United States to move forward on ambitious space plans, it will require the industry, executive branch, and legislature to work together to prioritize milestones. Every new technology or program cannot be the highest priority, which means that inside the executive branch some of these programs will have to be centrally managed while Congress exercises oversight and appropriations responsibilities. U.S. private industry also has a critical role since the United States, for some time, has encouraged civilian companies to compete for contracts related to space.

To be successful, private-sector space companies, NASA, the Department of Commerce, the Department of Defense, Congress, and the White House, among others, must all work together to unlock the limitless potential of space and America's potential in space. Table 8.1 summarizes the space-related missions of various government agencies.

CLARIFYING THE ROLE OF NASA

NASA has been the face of the U.S. space program for over half a century, serving America through its vision "to discover and expand knowledge for the benefit of humanity."[4] However, NASA's changing objectives and existing budget constraints have complicated long-term planning and resulted in debates about the agency's purpose and presence in space.[5]

Table 8.1   U.S. Government Agency Space Missions[i]

Department of Commerce	• "The Office of Space Commerce is the principal unit for space commerce policy activities within the Department of Commerce. Its mission is to foster the conditions for the economic growth and technological advancement of the U.S. commercial space industry."[ii]
Department of Energy	• "The mission of the Energy Department is to ensure America's security and prosperity by addressing its energy, environmental and nuclear challenges through transformative science and technology solutions."[iii]
Department of State	• "The Office of Space Affairs carries out diplomatic and public diplomacy efforts to strengthen American leadership in space exploration, applications, and commercialization by increasing understanding of, and support for, U.S. national space policies and programs and to encourage the foreign use of U.S. space capabilities, systems, and services."[iv]
Federal Aviation Administration	• "Safety is our North Star. We protect the public, property, and U.S. national security interests. We license all U.S. commercial launches and reentries here and abroad. We work with our international partners to promote safety. U.S. Spaceports are critical infrastructure for the industry."[v] • The "Office of Commercial Space Transportation (AST) was established in 1984."
National Aeronautics and Space Administration	• NASA's mission is to "explore . . . the unknown in air and space, innovate . . . for the benefit of humanity, and inspire . . . the world through discovery."[vi]
National Space Council	• "The National Space Council (NSpC) is the White House policy council responsible for ensuring the United States capitalizes on the rich opportunities presented by our nation's space activities."[vii]

[i] There are several other government agencies with space-related missions; this table only portrays summaries of the largest agencies with space missions.

[ii] U.S. Department of Commerce, Office of Space Commerce, https://www.space.commerce.gov/about/mission/.

[iii] U.S. Department of Energy, https://www.energy.gov/about-us.

[iv] U.S. Department of State, Office of Space Affairs, https://www.state.gov/bureaus-offices/under-secretary-for-economic-growth-energy-and-the-environment/bureau-of-oceans-and-international-environmental-and-scientific-affairs/office-of-space-and-advanced-technology/.

[v] Federal Aviation Administration, Commercial Space Transportation, https://www.faa.gov/space/.

[vi] "Our Mission and Values," NASA, https://www.nasa.gov/careers/our-mission-and-values.

[vii] The White House, National Space Council, https://www.whitehouse.gov/spacecouncil/.

The organization has to determine how best to spend a limited budget and whether funding should be prioritized for space exploration or terrestrial climate science. Pew Research polls on this subject are telling. When asked about NASA priorities, "majorities say monitoring climate or tracking asteroids should be a top NASA priority; only 13% say the same of putting astronauts on the Moon."[6] Moreover, a former deputy NASA administrator, Lori Garver, argues that there is no space race anymore and that humans face an existential threat of climate change, so resources are better spent addressing the risk to the environment rather than space exploration.[7]

Even when exploration is considered, there is an ongoing debate as to whether humans or robots should be exploring space.[8] Robots offer several advantages over living crews because there are no human safety concerns, and they are assembled in clean rooms to reduce the chance of any contamination on Mars's surface[9]

To summarize, recent polls show the U.S. public believes it makes sense to focus on the climate study of Earth rather than space exploration; there is debate that it may be safer to send machines instead of humans to space; and if we are landing on other celestial bodies, we have to be careful not to contaminate them. However, polls are strongly influenced by the priorities communicated by the government. As with other compromises, a reasonable desire to conserve our ability to learn about the origin of life must be balanced against the project to spread Earth life and intelligence to worlds now dead—it must not become an absolute precautionary principle or right of inanimate objects or microbes[10] that circumscribes the larger project of development and settlement that benefits the overall project of life and its continuity. A healthy concern with observing our environment and those of other planets is a necessary step to stewardship and planetary management,[11] and past polls have placed planetary stewardship (space solar power and planetary defense) at the apex of citizen interests.[12]

Communication to the American people is essential. It can be argued that NASA, or the U.S. government, has not relayed a rationale for why we should go back to the Moon or, for that matter, to Mars. Placing such projects purely within the realm of science is unlikely to be convincing. Support for exploration itself without a connection to a grander project as well as domestic and international agendas will continue to be anemic.

Over the past decade, the executive branch has vacillated over whether America should visit the Moon or Mars, if we should keep the International Space Station operational, and by what time frame specific milestones should be accomplished. The United States must also decide whether it wants a mere "flags and footprints" return to the Moon or a true, permanent, and productive space presence with settlement and commerce.[13]

The American people need a space vision to clarify these points and to set NASA and the rest of the government on the right course.

NASA has, and will continue to have, a secondary mission to carry out scientific research and conduct science diplomacy. However, if NASA is only considered a science agency, it would make more sense to merge it with the National Science Foundation and move its aeronautical and spaceflight functions to the Federal Aviation Administration, U.S. Air Force, and U.S. Space Force (USSF).

Yet NASA's primary mission is to advance U.S. interests through peaceful means and partnerships, which is often in direct competition with rivals. NASA was born out of a national security need for the United States to peacefully compete with the USSR in the Cold War. In the first space race, it was about winning the hearts and minds of the newly independent postcolonial states away from communism and toward the United States' vision of society by demonstrating technological supremacy. The focus was on firsts and prestige.

Today we also have a national security need to peacefully compete with China, but mere prestige and firsts are far less relevant—they are at best a sideshow, and NASA must not be distracted by the theatrics. Now, the need is to build and maintain a vibrant and competitive in-space economy that will be attractive to international partners as a vehicle for common prosperity. NASA's role here is to open new industries—extractive industries, in-space power, and in-space industrialization and manufacturing—and to enable the fullest extent of economic activity, including human pioneering and settlement. Because this mission is uncomfortable for many people at NASA who grew up during the period of U.S. hegemony, both the administration and Congress must give NASA strong, clear, and unequivocal direction to focus its efforts on advancing in-space industrialization—igniting and accelerating an off-Earth supply chain and manufacturing base.

A New Path Forward

Today's NASA is not called to be in the heroic limelight but to be a facilitator and organizer of the success of others—U.S. industry and U.S. citizens. Toward that end, Congress and the administration should encourage NASA to cede ownership and control of infrastructure and operations in low Earth orbit (LEO) to the private sector. Instead, NASA should position itself as a strong anchor tenant of commercial LEO space stations and champion U.S. commercial space stations as facilities where friendly nations can send astronauts and conduct science.

Because of its significant experience with planetary geology and deep spaceflight, NASA can aid in both prospecting for resources and assessing threats for the planetary defense mission, but Congress should give primary responsibility for planetary defense to the Department of Defense.

At present, because of SpaceX's strong ideological motivation—and the company's speed, significantly broader vision, and much lower cost to taxpayers—NASA should seek a Space Act agreement with SpaceX, supporting the firm's goals to establish a permanent and growing independent settlement on Mars. NASA should mature and make available tools such as nuclear rockets and propellant depots but should be dissuaded from seeking taxpayer dollars for a NASA-owned and -executed human Mars mission similar to its Lunar Exploration Program (which relies on the SLS rocket). As long as SpaceX is ideologically motivated, the company will be faster and more efficient than NASA, and U.S. tax dollars are better spent on NASA efforts aimed at speeding the creation of an Earth-independent manufacturing supply chain.

NASA has two basic tools to facilitate broader U.S. spacefaring: government buying power to enable, incentivize, and stabilize new markets; and government-sponsored precompetitive research to enable new industries.

In using its buying power, NASA should focus its commercial encouragement missions on areas where industry is ready or near ready to compete in a Commercial Orbital Transportation Services (COTS)-like program. The most urgent need is for NASA to develop a concept for a commercial-friendly Lunar industrial facility at the Moon's South Pole. It is important for NASA to seek a large number of sustainable commercial tenants. NASA must specify production targets to encourage scale and linked to broader national objectives such as space solar power and strategic minerals that can support a green economy. NASA should take the lead in supplying the initial basic infrastructure for the industrial park but should design it with open standards and an ability for industry to generate economies of scale. Through NASA, the United States could make available free infrastructure services to first entrants, analogous to land grants that incentivized the westward expansion.

In sponsoring precompetitive research, NASA should focus its technology development on technologies where industry is not yet ready to compete and where commercial investment is low, but where a significant economic advantage and scale could be achieved once the technological risk is eliminated. NASA's aim should be to speed commercial development. The most current critical need is to advance the fundamental technologies with respect to in-situ resource utilization (ISRU) technologies related to in-space manufacturing and extraction of materials from the Moon and asteroids.

PRIVATE-SECTOR PARTNERSHIPS

While NASA is often at the forefront of space policy, it is no longer the unheralded leader. America is fortunate to have a free society where individuals with an entrepreneurial spirit are free to achieve their dreams. Musk, Bezos, Branson, and others are the most dynamic driving forces behind U.S. advancements in space. The thriving space private sector would not be possible without the opportunities enabled by public policy decisions, and it is important that policymaking does not provide road-blocks as it has in the past but, rather, works toward a successful symbiotic private-public partnership in space.

Silicon Valley versus Washington, DC

One of the more complicated problems confronting the formulation of space policy is determining how best to allow private-sector companies to support government space services, particularly those with unique technological prowess. A disconnect currently exists between Silicon Valley and Washington that prohibits talented engineering and tech companies from contributing more meaningfully to government projects. SpaceX provides a good example of a Silicon Valley company that was able to develop impressive rocket technology and—through perseverance, strong financial backing and, ultimately, the will to move forward—successfully overcome many barriers to secure launch contracts with the U.S. government. The path to success was not easy, however, and SpaceX had to wage a multiyear legal battle to convince the U.S. Air Force to end the monopoly of existing defense contractors and purchase SpaceX technology.[14] How many other companies would be willing to work with the government if it was easier for them to gain access? As it stands, achieving the "unicorn" status of government space service contractor is extremely difficult, and more often than not, private firms simply bypass projects in the defense sector because it is not profitable or worth the effort.[15]

The divide between Silicon Valley and Washington can be explained most simply by economics. As Rachel Olney outlined in War on the Rocks, "Defense contracting is a risky business for fast-moving commercial companies, especially startups. Without aligned financial incentives, high-tech companies cannot pursue the national security market and therefore, many state-of-the-art commercial technologies will never reach the warfighter."[16] Moreover, some companies are morally hesitant to work with the Pentagon. Yet these issues must be overcome to attain top-end talent to support future space initiatives.[17] It is particularly critical to do so, moreover, because—while disagreements may exist between Washington and Silicon Valley—the Chinese Communist Party (CCP) has embraced a strategy of military-civil fusion[18] and is directing a whole-of-government effort toward space development and with no shortage of talented labor to fuel its space initiatives.

Talent Management and the Technical Base

In addition to focusing on how the Department of Defense can partner with the private sector from a macro perspective, the United States must also pay attention to building a domestic future-oriented space workforce. The emerging space sector provides an economic opportunity for new jobs in several technical sectors, including engineering, computer security, artificial intelligence, energy, and mining, among others.[19] The United States will need to find ways to foster a space innovation ecosystem, similar to the cyber hub model seen in Israel and England. Developing and training a workforce will be essential to ensuring that employees are available for the burgeoning space sector—this will begin with inspiring the next generation of science, technology, engineering, and math (STEM) leaders. According to a recent report that convened large swaths of the space industrial base and government sector, the U.S. government will need to fill 10,000 STEM jobs for the NASA Artemis program alone in the very near future.[20] The government will have to emphasize retaining top U.S. citizens capable of performing in these new space careers.[21] Unfortunately, the trend lines are not favorable, as China is on pace to have nearly double the number of STEM doctoral graduates by 2025 compared to the United States,[22] and the scientific workforce in the U.S. government has already been constricting.[23] Steps will need to be taken immediately to correct these trends.

EMPOWERING THE DEPARTMENT OF DEFENSE

With enabling policy, the U.S. space economy can grow to a significant scale, contributing to the nation's economic security and providing the war chest and industrial might needed to serve as a deterrent as well as the material capabilities needed to win in a military contest.

However, for the United States to prevail in shaping the contest within the space area of responsibility and in a terrestrial conflict, the Department of Defense and the USSF, in particular, must evolve.

Maximizing U.S. Spacepower

A space force is much more like a navy than an air force. The U.S. Navy certainly supports the joint forces by moving troops and cargo and flying air support missions—but that is not its main mission, nor why it was created. Similarly, the USSF will support the joint forces through a variety of space-based information services like precision navigation and timing; missile warning; over-the-horizon communications; and tactical overhead intelligence, surveillance, and reconnaissance—but that is not its main job nor why the service's supporters lobbied for its independence from the U.S. Air Force. Legacy military missions could be accomplished through its predecessor, the Air Force Space Command, but the USSF was created to enable a broader mission set. The U.S. Navy certainly has the mission and ability to execute independent options, whether strike or blockade, to coerce a nation—but that is a derivative mission. The USSF also must have liberty to pursue independent options within and from the domain.

But the U.S. Navy was created to secure commerce and access to distant sources of wealth—it was authorized by Congress to stop the Barbary States from preying on American merchant ships. As the great naval theorist Alfred Thayer Mahan stated, "The necessity of a navy . . . springs, therefore, from the existence of a peaceful shipping, and disappears with it. . . . When for any reason sea trade is again found to pay, a large enough shipping interest will reappear to compel the revival of the war fleet. . . . Naval Strategy has for its end to found, support, and increase, as well in peace as in war, the sea power of a country."[24]

So too, the USSF has a critical role to extend the spacepower of our nation in peace as well as in war, and its fundamental necessity rises from our expectations of expanding commercial interests and our need to proactively support and secure our expanding space lines of communication. This is a daily presence mission to build up the domain, secure the important logistical nodes and routes, and build up a mighty space industrial base.

A key relationship exists between the United States' commercial space industry and the USSF. The United States currently spends 3.7 percent of gross domestic product on defense. If we were to allocate that proportionally by domain, then the U.S. space economy, estimated at about $200 billion, would legitimately have claim to about one-third of the USSF's present budget for space security and defense. But over time, this share will grow significantly, eventually first equaling and eclipsing the terrestrial war-fighting component of the budget around 2030 and then likely becoming double its size by about 2045. A space force committed to the "defense of commerce" and the space security mission will have a legitimate claim to that budget, but a space force that eschews this responsibility does not.

A space force that maximizes U.S. spacepower is jealous of its missions. It should seek as much as possible and appropriate to be a one-stop shop for required action in the space domain, similar to how the U.S. Coast Guard is for the sea. It should actively seek space security missions such as planetary defense; active debris removal; and authorities for vessel search, inspection, board, seizure, and law enforcement.[25] It should have a proactive legislative agenda to seek authorities for maximum freedom of action such as multiyear acquisition, the authority to issue infrastructure bonds, the authority to maintain an in-space strategic reserve, its own working capital fund, and, of course, its own department and secretary free from air allegiances.

Shaping the Department of Defense for Space Prowess

To set the Department of Defense on a path for success in space, leadership can take several steps to proactively advance space missions. First, the department needs to have clarity of mission and know its role. To achieve this objective, the Office of the Secretary of Defense should codify key space security roles and missions of the USSF in Department of Defense Directive 5100.01, which outlines the major functions of the department.[26] These should include defense of commerce, planetary defense, and military support to civilian authorities for law enforcement in space.

Along those same lines, civilian leadership should provide necessary focus to U.S. Space Command and USSF planning efforts that these organizations are currently lacking. The Guidance for the Employment of the Force should task USSPACECOM with a campaign plan and shaping end states, including the safety and security of the homeland from the threat of natural potentially hazardous objects (asteroids and comets). More specifically, USSPACECOM should be tasked with concept plans in the Joint Strategic Capabilities Plan for forward-thinking scenarios for military operations other than war, such as planetary defense against asteroids, rescue of U.S. citizens in space, noncombatant evacuation, and vessel search and seizure, as well as war-winning plans.

With a clear mission in place, next it will be imperative to have a trained force capable of executing these objectives—this prompts the need for a strong professional military education (PME). PME efforts must include cognizance of the ongoing great power competition and must include deep study of the benefits for the USSF that accrue from building an in-space industrial base for national enduring advantage.

In addition to improved personnel, key departments must be expanded to fill new roles. The Defense Advanced Research Projects Agency (DARPA) is a prime candidate to contribute to U.S. space capabilities. DARPA exists to prevent technological surprise for the United States and to cause such surprise for its opponents. The agency has a number of subordinate offices but none devoted to space and the development of industrial space capabilities. The secretary of defense should direct DARPA to set up an exoplanetary office to mature high-risk, disruptive, and revolutionary investments that enable space industrialization.

Finally, with purpose, planning, people, and programs in place, the Defense Department will also require predictability—particularly with launch services—to have a greater impact in space. With the ready availability of low-cost commercial launch, generic satellite busses with plug-and-play payload interfaces, the ability to mass-produce low-cost space hardware, and a national direction moving toward in-space refueling, the department has an opportunity to move part of its requirements to a first-in, first-out warehousing and launch-on-schedule responsive launch capability. Moving to launch on schedule, especially with known standard satellite busses, enables a predictable and rapid cadence of experiments and access to space that is closer to the special operations command model, where schedule and price are fixed and performance is the best available.

When nothing is ready to launch, the Defense Department should launch propellant to create prepositioned operational stores similar—but not identical—to the department's War Reserve Materiel program, enabling a maneuver advantage and allowing new launch entrants to space qualify and insure their vehicles.

CONGRESS: LEGISLATING FOR SPACE PRIMACY

Congress has immense power to shape the incentives and purposeful direction of the bureaucracy that executes and enables U.S. spacefaring. Clear direction is what is required, equal to the nation's ambition. These, in turn, affect the incentives and liberty that U.S. companies and citizens can use to find opportunity and prosperity in the space domain.

The first measure that Congress must undertake is to codify the national purpose for space—there may not be a more important action. This purpose must be clearly stated in the authorizations for both NASA and the Defense Department.

Next, Congress will need to show individuals and businesses the value of space by clarifying the rights of U.S. citizens in space. U.S. space law expert Wayne White has provided a model act, the Space Pioneer Act, that provides the level of clarity needed to incentivize individual businesses and citizens, similar to the Homestead Act of 1862. This model, or a closely related substitute, must be clearly enacted into law.

In addition to demonstrating how U.S. private corporations and individuals can benefit from space, it is necessary to take steps to bolster the space economy by creating clearly labeled funding lines for in-space servicing, assembly, and maintenance (ISAM) and ISRU programs in the budgets for both NASA and the Defense Department.

Fostering and Financing Space Initiatives

ISAM and ISRU will be integral parts of the emerging space infrastructure, but continued growth is not possible without proper government-enabled funding measures in place. Congress will need to create infrastructure-enabling authorities to incentivize and spur development of infrastructure that will power an in-space economy. There are several financial vehicles that can be deployed to boost space economic development.

For one, issuing bonds may increase investor interest in space. On Earth, we do not expect to fund major infrastructure projects solely on annual appropriations. Rather, federal, state, and municipal governments authorize and issue multiyear bonds to fund necessary infrastructure improvements. Congress should enable one or more agencies or departments (e.g., the Department of Commerce, NASA, or the USSF) to issue space bonds for space infrastructure.

Along these lines, "public capitalization notes" may be another option. Novel vehicles exist to directly incentivize the creation of new value. Economist Armen Papazian has developed the concept of public capitalization notes specifically to enable the economic development of space.[27]

A third option is the creation of a space infrastructure public corporation. While NASA explores and conducts research in, and the USSF defends, the space domain, whose job is it to actually encourage and develop space-based public infrastructure? A new organization is needed to fill this critical gap. The Foundation for the Future (F4F) has provided a model for how to set up a public corporation with a focus on infrastructure for space economic development.[28]

Another option is to create a space commodities exchange. On Earth, commodities markets are enabled and stabilized by commodities and futures commodities exchanges and are regulated in the United States through the Commodities and Futures Trading Commission. Economist Bruce Cahan has developed a "shovel-ready" proposal to create a space commodities exchange that can provide the essential market-making infrastructure for space resource development and recovery.[29]

Government seeding of those commodities will be required to prime the pump of the materials and propellant that will fuel the space economy. This can be achieved by authorization of an in-space national strategic reserve for minerals and propellant. Congress should authorize (and designate a lead—for instance, Defense, Commerce, or NASA) for a national strategic reserve in space that would purchase and collect space-sourced propellant and strategic minerals in Cislunar space (such as at the first Earth-Moon Lagrange point—EML1[30]). Using government's buying power would create an initial market to ignite a self-sustaining market for space commodities.

Finally, even with energy measures in place, development will not occur instantaneously. One financial tool that should be considered is multiyear funding for robust international collaboration. In defense sales, U.S. and partner appropriations are kept in interest-bearing U.S. Treasury accounts for as long as seven years. International science and technology cooperative agreements and cooperative space infrastructure projects take years to negotiate and advance security cooperation and should be treated similarly. Congress should authorize all international space collaboration to have the same fiscal latitude available for foreign military sales—this would be enabling for both the Defense Department and NASA.[31]

Space Power and People

Energy is a key investment for space development. In-space power and advanced propulsion are the fundamental enablers of America's space ambitions. While many aspects of the emerging space economy are well capitalized with multiple commercial actors, power and propulsion face significant barriers to commercial entry because of the longer timelines to develop profitable products (outside normal investment timelines), the substantial infrastructure required to develop these products, the significant regulatory hurdles to production, and the high level of investments required to complete finished products. Today, power and propulsion are not "Moore's Law technologies"—where expectations and market forces lead to regular doublings of capability. They require work and investment to retire technical risk. Advanced power (greater than 1 megawatt, or the amount of energy required to power 400–900 American homes for a year) and advanced propulsion (twice the performance efficiency of chemical propulsion) programs should be funded at no less than $1 billion each in the NASA and USSF budgets.

Along with power, you need people. The ability of the United States to lead the emerging in-space economy depends critically on its talent base, especially in STEM fields and not only in higher education but also in blue-collar jobs. The talent resources of America's competitors are so substantial that, to compete, America must mobilize and unlock the full human talent available to it across all space disciplines, including craftsmanship.[32] Just as the United States provided national security grants after Sputnik that fueled innovation for generations, it needs new scholarship and payback opportunities to build technical and production acumen.

Associated Agency Space Authorizations

Although not at the forefront of space policy, the departments of Commerce, State, and Transportation all play a role in actively shaping American interests in space. Congress has an opportunity to provide clear direction to help enhance space policy through these entities.

As the space economy grows, the Commerce Department will play an outsized role in its development. The nation needs a more powerful advocate for commercial space within Commerce. Congress should elevate the Office of Space Commerce—which is currently buried in the Commerce Department within NOAA's National Environmental Satellite, Data, and Information Service (NESDIS)—to become its own bureau. Given that the Biden administration's proposed budget for fiscal year 2023 has included funding for this,[33] Congress should speed this enabler for American prosperity.

Congress should also provide funding and a tasking for Commerce to conduct a decadal study to inform U.S. government investments in space industry. Decadal studies are considered the gold standard of scientific community goal setting and advice to government,[34] but no such collaborative priority setting has been accomplished for space industry investments. (The National Academies of Sciences and Engineering serve as a good reference model for successful decadal studies.) The Commerce Department is a more appropriate customer than NASA or the Defense Department, though the study should make recommendations for NASA, the Defense Department, and the Department of Energy.

In addition to focusing on the Commerce Department, policymakers should also provide guidance to the Department of Transportation. Congress should continue its space agency transformation by formally elevating space to be a full mode of transportation at the department.

As the United States works to build space initiatives on the home front, the State Department should simultaneously be working on international space policy. While the Artemis Accords currently being advocated by the State Department is a great start to securing a pro-development reading of the Outer Space Treaty in international law, more work is required.[35] Congress should task the State Department to develop and brief to Congress a proposal to develop a suite of new international organizations to enable the space economy. Similar to the post–World War II institutions that allowed America to shape the postwar order, the United States should seek to create a series of institutions likely to promote U.S. preferences. The required models include an International Civil Spacefaring Organization modeled on the International Civil Aviation Organization, an Organization for Economic Development in Space modeled after the Organization for Economic Cooperation and Development, a Space Development Bank analogous to the World Bank or International Monetary Fund, and a Space Security Collective Defense Community analogous to the North Atlantic Treaty Organization.

New Department of Defense Directives

The Department of Defense took a major step forward with the creation of the Space Force, but it must be only the first step of many to serve a much-needed role in the American space enterprise. Congress provided broad authority to protect U.S. interests in space to both USSPACECOM and the USSF; however, there is a lack of clarity about implied tasks that are critical to the safety and security of U.S. spacefaring. At a minimum, explicit legislation should task the Defense Department with defense of commerce in space and with planetary defense.

While the U.S. Navy is generally thought of as the analogue for the USSF regarding the role of commerce protection, the USSF should be granted U.S. Coast Guard–like authorizations in space. Congress should strongly encourage extending Title 14 U.S. Coast Guard–type authorities that would enable the USSF to execute law enforcement missions in space. The Coast Guard is an excellent model for the Space Force, given its example of one-stop-shop governance of American maritime interests. It is simultaneously a military and law enforcement agency. The United

States dual hats other agencies, such as in the cyber domain, to enable flexible responses to emerging situations. The United States is expecting a significant increase in commercial activity in space, some of which could flout U.S. laws and regulations, cause international incidents, or involve illicit activity. For the foreseeable future, the USSF will be the only U.S. government actor capable of space inspection and enforcement. Moreover, the USSF will have to deal with competitors who are likely to use gray-zone tactics they use on Earth with a full range of authorities—the USSF should have every authority it takes to compete and the ability to manage the totality of U.S. security interests.

Space military capabilities will be nothing without funding. Major "customers" of USSF capabilities (primarily, the other terrestrial armed services) deserve a say in both how effort is allocated toward space support services and the kind of systems purchased. On Earth, we have found it valuable to create an artificial market to efficiently allocate military airlift, called the Transportation Working Capital Fund (TWCF). A dedicated Working Capital Fund for Space would allow clear demand signals from other military units to vote with their dollars as well as to keep the core space budget focused on building enduring advantage. It would enable efficiencies for government purchases of commercial space services through a single portal. Virtually anything can be conceived as a service—satellite communication as a service, launch as a service, infrastructure as a service, computation as a service, refueling as a service, space domain awareness as a service—all of which could allow the USSF to use government buying power to build enduring advantage by igniting and sustaining a vibrant and innovative space industrial base. A Space Working Capital Fund should allow other Defense Department and U.S. government customers to "purchase" services from the USSF, as well as allow the USSF to use this fund to purchase commercial services. Of note, TWCF authorities even allow non–U.S. government customers to use military airlift.[36] Congress should allow the USSF to provide services for authorized non–U.S. government customers on a reimbursable and non-distracting basis because opportunities will exist to use USSF capabilities to advance U.S. interests (e.g., rescue, towing, inspection, refueling, and spacelift).

For the USSF to move at speed to take full advantage of commercial services, its fledgling Commercial Services Office[37] must expand in both its resources and its authorities. Today the office has been asked to stay away from launch so as to protect incumbent interests both within and without the government and so as to not be a disrupter like the Space Development Agency (SDA)—though the SDA has *constructively disrupted* the slow pace of satellite procurement. Purchasing launch as a service would certainly be constructively disruptive (as the Defense Innovation Unit has demonstrated).[38] The office should be given broad latitude to maximize anything "space as a service." It should also be given a significant budget. Various subject matter experts contributing to the "State of the Space Industrial Base" reports judged the appropriate budget to be $1 billion.[39]

The USSF will not be effective as a software-only digital service. It will require dedicated military hardware and at times need to rely on civil crafts to augment its capabilities. It is highly likely that the civil fleet of spacecraft (satellites, launch vehicles, and in-space transport vehicles) will expand in capacity much faster than military systems. On Earth, while the Air Force does maintain a military airlift fleet, it augments this with a Civil Reserve Air Fleet (CRAF), which allows for contract airlift in peacetime and mobilization in times of national emergency.[40] The CRAF sets standards[41] critical to support military airlift and provides some subsidy for the increased costs. A CRAF-like model would allow the USSF to take advantage of civil innovation and efficiencies and be ready to supplement the nation in times of emergency. Congress should not wait for an emergency to create such an authority but should lay the groundwork now to allow the USSF maximum flexibility to shape and leverage commercial space fleets.

As Congress works to hone the role of the USSF and define responsibilities during peacetime and war, it should consider the big picture and increase the prominence of the USSF commensurate with its vast and growing responsibilities in the largest domain. Forming a separate service may be the biggest step yet, but Congress should further extend the USSF's independence by granting the USSF its own separate department with its own separate secretary at the nearest opportunity. The nation is not served by keeping the Space Force shackled within the Department of the Air Force. No intrinsic link exists between the air and space domains. Air and space are distinct physical and maneuver domains requiring different expertise. Air and space are distinct legal domains. Air and space combined arms are no more closely related than space and land, space and sea, or space and cyber. The USSF should not remain a captured organic support arm of the Air Force (it continues to be), because it's necessary role in securing America's space supremacy is much bigger. The nation will benefit when the Space Force is free to make the partnerships that make sense with the Army, Navy, Marine Corps, and Coast Guard without the coercive power of a single service (the Air Force) to control its access to key support services, such as legislative liaison, recruiting, and budgeting processes.

The space domain is undergoing unprecedented growth and change and deserves the full attention of a dedicated secretary. The space domain offers the United States the potential to realize trillions of dollars in economic impact, and potentially millions of jobs for its citizens, if space is secured and nurtured by the Space Force. No similar potential exists for the Air Force to advance the nation's competitive strength through aviation. Our nation's future in space is important; the secretary of the Air Force's energies should not be distracted by the management of legacy and wasting air assets nor overpowered by the considerations of the air domain because the department's senior service has a budget seven times larger, or 55 times the number of personnel to choke the Air Force secretary's office with their own problems. America meekly took a halfway step to help get the Space Force on its feet. Now that the USSF has its footing, it is time to remove the shackles and let it control its own destiny. If the United States is serious about winning the future in space, it must organize for victory, not mere bureaucratic status quo–assuaging convenience. Congress should create a Department of the Space Force.

Furthermore, building on that new department, the USSF should establish a real and robust Space Force Reserve component. The National Guard and reserve systems established for the terrestrial services support the United States by enabling expertise and talent to be retained, refreshing the military with modern business practices, and allowing service for those who wish to serve part time or to put down permanent roots in a community. We put people in uniform when they must act under authority to kill or destroy or when they must be ordered directly into harm's way. We use civilian and contract personnel who are not part of the accountability related to lethal orders when they are needed to augment the technical and managerial experience of active-duty troops. An active-duty force exists to stand ready at all times, to be fully deployable. A reserve component allows part-time and flexible scaling with demand. A National Guard component allows in-place homesteading of missions that do not need to move, that are connected with homeland defense in their communities, and that have synergies with emergency response and economic development at the state level.[42] It is wholly unacceptable for a significant portion of our Guardians to be orphaned—left in other services, even if "and Space" is taped to the Air Force National Guard's name—and unable to be formally part of the Space Force. It is equally unacceptable to allow reserve Guardians to become second-class citizens as part-time employees in a monolithic "single component" Space Force monopolized by active-duty interests as the Defense Department and Department of the Air Force are currently imposing. Congress should act promptly to establish both reserve and National Guard components, even in the face of Defense Department resistance.

Yet another consolidation that must be considered relates to what other parts of the U.S. national security space enterprise should exist under a Department of the Space Force. A key motivation for the creation of the USSF was to reduce the number of players to enable central management. The most significant outstanding parts are the National Reconnaissance Office (NRO) and the Missile Defense Agency (MDA). The NRO, which already uses Space Force personnel, has acquisition authorities that could be aligned under a secretary of the USSF, maintaining its independence and authorities, not unlike the SDA. Operationally, the NRO could become one or more Strategic ISR Deltas, which operationally report to USSPACE-COM. The MDA cooperates well at present, but if the Space Force is given a formal missile defense mission, the space portion of the MDA should be transferred to the USSF.

Finally, the USSF would not be complete without some internal designers and builders to help develop the space infrastructure. Developing space is likely to require significant public works and some level of regulation. On Earth, the U.S. Army Corps of Engineers has special authority to build and maintain public works (such as canals, dams, and levies) and to regulate and maintain certain public works. Congress should similarly empower the USSF with its own Space Corps of Engineers with the authority to construct public works on the frontier.

RECOMMENDATIONS FOR THE PRESIDENT AND THE NATIONAL SPACE COUNCIL

The two most recent U.S. presidential administrations have shown a decided interest in advocating for space and have revitalized the role that the government plays in shaping U.S. space policy. The reconstituted National Space Council (NSpC) has a particularly important responsibility to greatly affect American advancements in space.

Similar to Congress, the NSpC needs to focus on bolstering space development. The in-space servicing, assembly, and manufacturing strategy[43] is an excellent start, but it lacks achievable goals, benchmarks, and timelines caged to America's competitors and the fastest speed at which industry can move. The next step to foster developments in space is to publish implementing ISAM guidelines.

The NSpC should draft an executive order for presidential signature giving clear guidance to agencies, including the Office of Management and Budget, to ensure unity of effort and measurable milestones. It will be equally vital to focus on extractive industries and power, so the NSpC should follow up on its ISAM strategy with a companion strategy for "space mining" or ISRU and a companion strategy for space-based solar power, providing a visionary and scalable application of ISAM applied to defeating climate change.

Part of implementation should be to elevate various ISAM and ISRU components for a Defense Production Act (DPA) Title III Presidential Determination.[44] These space activities are essential as the capabilities represent industrial resource, material, or critical technology items essential to national defense, making them eligible for the DPA Title III authorities to create, maintain, protect, expand, or restore domestic industrial base, and making available loan guarantees, advance purchase commitments, and installation of equipment for production.[45]

Along those same lines of metric-based measured progress, guidance for the Moon is also important. One way to guarantee progress is for the NSpC to set forth production targets for the NASA-led public-private partnership Lunar industrial facility at the Moon's South Pole. These production targets should be chosen to catalyze industry to create scalable solutions for regolith extraction, processing, and production of commodities for propellant, oxygen, metals, and glasses. Initial targets should start in the hundreds or thousands of metric tons and grow from there.

Finally, the NSpC should work to prevent the threat of extinction-level events by publishing an executive order on planetary defense. The NSpC needs to delineate the respective roles and responsibilities of the Defense Department, NASA, the Energy Department, and the Department of Homeland Security/Federal Emergency Management Agency. It should also make certain planetary defense is mentioned in the Unified Command Plan and the Guidance for the Employment of the Force. Since 2008, no administration has responded to Congress's request to identify an agency responsible for planetary defense.[46] USSPACECOM clearly sees this as an implied task and has the operational role of dealing with threats in the space area of responsibility.[47]

Additionally, the president and Congress must seriously consider elevating the NSpC from an executive branch working group subject to the whims of the serving administration's or the vice president's interest in space to a legislatively constituted, permanent, and empowered governing board to design and execute the American space program. Establishing such a board, with the mission of steering the whole-of-government unity of effort for American space primacy, will become increasingly necessary as other organizational advances occur, including a separate Department of the Space Force. The years of history and service behind the NSpC make the council a ready-made solution to governing the new American space program as it increases in importance and complexity.

Moreover, the NSpC should have an organization, such as a presidential task force,[48] to monitor daily execution of NSpC intent. The right location for this task force is within a USSPACECOM joint interagency coordination group.[49]

An intriguing alternative that is worthy of debate, proposed by Dr. Brent Ziarnick along the British Admiralty model, would be a cabinet-level Department of Space that would take over management of NASA, the USSF, USSPACECOM, and a space infrastructure development organization or space economic development authority.

ARTICULATING AN AMERICAN SPACE VISION

Government, academia, and industry came together to support U.S. primacy in space during the Cold War. But the United States has ceded this high ground since 1990, and China is now on the verge of delivering

America another Sputnik moment—one that we will not be able to over-come without taking timely action.

The way to galvanize a populace is through a national vision, and China has already issued its call to action. The CCP propaganda machine has been in full effect for years, and its people have been spoon-fed the narrative of "national rejuvenation"—that they will ascend to reclaim their position of international prominence.[50] Likewise, there have been times in recent U.S. history that demonstrated the benefits of a well-articulated vision, so the importance and potential impact of such a vision for space cannot be understated.

To create an enduring vision, we recommend Congress engage in a series of hearings to examine the use cases of space. The recent RAND Europe "Future Uses of Space Out to 2050" report provides a useful outline of the big ideas the American public should be exposed to.[51] Engaging and educating the U.S. public on the importance of space and the benefits to the U.S. economy is likely one of the most essential undertakings to guarantee a prosperous future in space.

Above all else, the United States should aspire to be a spacefaring civilization—to extend the blessings of liberty to ourselves and our posterity, to encourage the outward expansion, economic development, and settlement of space under American values of freedom and human rights for the benefit of all humankind. To achieve this vision, Congress, the Defense Department, NASA, and the private sector should follow the guidance and directives addressed herein to focus on advancing the essential space activities that provide strategic and economic advantage, including space solar and nuclear power, in-space manufacturing and resource extraction, space-based tracking technologies, and organizational innovation to ensure U.S. superiority in space in perpetuity.

The window is closing for U.S. leadership to implement the vision for the next great strategic frontier. America certainly has the ability, ingenuity, and perseverance to rise to the occasion when challenged, but the time to act is now—ad astra.

Chapter Highlights

- **Clarifying the role of NASA.** NASA should be a strong anchor tenant of commercial space stations, cede ownership and control of infrastructure and operations in low Earth orbit to the private sector, and advance the fundamental technologies (in-situ resource utilization technologies and in-space manufacturing and extraction of materials from the Moon and asteroids).

- **Private-sector partnerships.** The Department of Defense will need to engage and incorporate talented Silicon Valley space startups and provide opportunities to develop and maintain a workforce capable of supporting space programs.

- **Empowering the Department of Defense.** The U.S. Space Force has a critical role in extending U.S. spacepower by expanding and protecting commercial interests; building up the domain; securing important logistical nodes and routes; and bolstering the space industrial base, planetary defense, active debris removal, and authorities for vessel search, inspection, board, seizure, and law enforcement. A separate Department of Space Force should also be created.

- **Congress: Legislating for space supremacy.** Congress needs to bolster the space economy by creating clearly labeled funding lines for ISAM and ISRU.

- **Fostering and financing space initiatives.** Potential vehicles to aid development include space bonds, public capitalization notes, a space infrastructure development public corporation, a space commodities exchange, an in-space national strategic reserve for minerals and propellant, and multiyear funding for international collaboration.

- **Executive space leadership.** The National Space Council needs to focus on bolstering space development, space solar power, and extractive space industries and establishing metric-based milestones for space.

- **Articulating an American space vision.** The United States aspires to be a spacefaring civilization; to extend the blessings of liberty to ourselves and our posterity; and to encourage the outward expansion, economic development, and settlement of space under American values of freedom and human rights for the benefit of all humankind.

Appendix
Space Strategy Podcast Guests

Episode	Podcast Guest	Podcast Guest Title and Organization
1	Mr. Richard Harrison	Vice president of Operations and Space Policy Initiative codirector, American Foreign Policy Council (AFPC)
2	Mr. Coen Williams	Director of Outreach for the 602 Club, a grassroots professional military association for junior Space Force Guardians and military space professionals
3	Mr. Douglas Loverro	President of Loverro Consulting, former associate administrator for Human Exploration and Operations at NASA, and former deputy assistant secretary of Defense for Space Policy
4	Mr. Rick Tumlinson	Founder of SpaceFund
5	Mr. Robert Zubrin	Founder of the Mars Society
6	Dr. Brent Ziarnick	Spacepower theorist
7	Mr. John Mankins	Founder of Mankins Space Technology Inc.

(continued)

Episode	Podcast Guest	Podcast Guest Title and Organization
8	Ms. Kara Cunzeman	Senior project leader for Strategic Foresight for the Center for Space Policy and Strategy at the Aerospace Corporation
9	Mr. Joshua Carlson	Author of *Spacepower Ascendant*
10	Brig. Gen. Steven "Bucky" Butow	Director of the Space Portfolio at the Defense Innovation Unit
11	Ms. Oriana Skylar Mastro	Center fellow at Stanford University's Freeman Spogli Institute for International Studies
12	Mr. Joel Sercel	Founder and CEO of TransAstra Corporation
13	Gen. James Cartwright	Harold Brown Chair in Defense Policy Studies for the Center for Strategic and International Studies, former commander of United States Strategic Command (USSTRATCOM), and former vice chairman of the Joint Chiefs of Staff
14	Dr. M.V. "Coyote" Smith, Col., USAF (Ret.)	Leading intellectual father of the Space Force, spacepower advocate, and professor of strategic studies at the Air University Department of Spacepower's Schriever Space Scholars program
15	Mr. Bill Bruner	CEO and cofounder of New Frontier Aerospace
16	Ms. Victoria Samson	Washington Office director for the Secure World Foundation
17	Ms. Mandy Vaughn	Founder and CEO of GXO
18	Dr. Bhavya Lal	Senior adviser to the NASA administrator on Budget and Finance
19	Mr. Tyler Bates	Captain in the U.S. Space Force
20	Col. Eric Felt	Director of the Air Force Research Lab Space Vehicles Directorate
21	Dr. Scott Pace	Director of the Space Policy Institute at George Washington University's Elliott School of International Affairs
22	Ms. Theresa Hitchens	Space reporter for Breaking Defense
23	Mr. Tim Chrisman	Cofounder and executive director of the Foundation for the Future (F4F)

Episode	Podcast Guest	Podcast Guest Title and Organization
24	Brig. Gen. John Olson	Mobilization assistant to the chief of Space Operations, Headquarters U.S. Space Force, and the chief data and artificial intelligence officer for the Department of the Air Force, Pentagon, Washington, DC
25	Mr. George Pullen	Chief economist of Milky Way Economy, instructor in Space Economics at Columbia University, and senior economist of the U.S. Commodity Futures Trading Commission
26	Brig. Gen. Steven "Bucky" Butow	Director of the Space Portfolio at the Defense Innovation Unit
27	Mr. Todd Harrison	Director of Defense Budget Analysis, director of the Aerospace Security Project, and senior fellow with the International Security Program at the Center for Strategic and International Studies
28	Mr. Kevin O'Connell	Founder and CEO of Space Economy Rising
29	Mr. Dean Cheng	Senior research fellow with the Heritage Foundation on Chinese political and security affairs
30	Mr. Wayne White	Author of the Space Pioneer Act
31	Mr. Jean-Jacques Tortora	Director of the European Space Policy Institute in Vienna
32	Mr. Henry Sokolski	Executive director of the Nonproliferation Policy Education Center
33	Mr. Dennis Wingo	CEO of SkyCorp Inc.
34	Mr. Howard Bloom	Best-selling author and founder of the Space Development Steering Committee
35	Lt. Gen. John Shaw	Deputy commander of the U.S. Space Command

Notes

CHAPTER 1

1. For more on NASA spin-offs, see Marianne J. Dyson, "Spacesuits, Firefighters, and Helping Heroes," National Space Society, *Ad Astra* 13, no. 6 (2001), https://space.nss.org/spacesuits-firefighters-and-helping-heroes/; and NASA Spinoff, https://spinoff.nasa.gov.

2. Thomas G. Roberts, "Space Launch to Low Earth Orbit: How Much Does It Cost?" Aerospace Security Project, Center for Strategic and International Studies, September 2, 2020, https://aerospace.csis.org/data/space-launch-to-low-earth-orbit-how-much-does-it-cost/; see also "Space: The Dawn of a New Age," Citi GPS: Global Perspectives & Solutions, May 2022, https://ir.citi.com/gps/kdhSE NV4r6W%2BZfP44EmqY4zHu%2BDy0vMIZnLqk4CrvkaSl1RIJ943g%2FrFEnNLi T1jB%2BjLJV4P9JM%3D.

3. The White House, "United States Space Priorities Framework," December 2021, https://www.whitehouse.gov/wp-content/uploads/2021/12/United-States -Space-Priorities-Framework-_-December-1-2021.pdf.

4. Christian Davenport, "Investors Are Placing Big Bets on a Growing Space Economy. But Can They Reach Orbit?" *Washington Post*, September 5, 2021, https://www .washingtonpost.com/technology/2021/09/05/space-finance-bubble-investors/.

5. "Space: Investing in the Final Frontier," Morgan Stanley, November 7, 2018, https://www.morganstanley.com/ideas/investing-in-space.

6. "Space: The Dawn of a New Age," 22.

7. Brian Higginbotham, "The Space Economy: An Industry Takes Off," U.S. Chamber of Commerce, November 11, 2018, https://www.uschamber.com/series /above-the-fold/the-space-economy-industry-takes.

8. Michael Sheetz, "The Space Industry Will Be Worth Nearly $3 Trillion in 30 Years, Bank of America Predicts," CNBC, October 31, 2017, https://www.cnbc

.com/2017/10/31/the-space-industry-will-be-worth-nearly-3-trillion-in-30-years
-bank-of-america-predicts.html.

9. Wilbur Ross, "A New Space Race: Getting to the Trillion-Dollar Space Economy World Economic Forum, Davos, Switzerland," U.S. Department of Commerce, January 24, 2020, https://2017-2021.commerce.gov/news/speeches/2020/01/remarks-secretary-commerce-wilbur-ross-new-space-race-getting-trillion-dollar.html.

10. The White House, "Remarks by Vice President Pence at the Fifth Meeting of the National Space Council | Huntsville, AL," March 26, 2019, https://trumpwhite house.archives.gov/briefings-statements/remarks-vice-president-pence-fifth-meeting-national-space-council-huntsville-al/.

11. Kamala Harris, "Remarks by the Vice President During a Visit to NASA Goddard Space Flight Center in Greenbelt, Maryland," American Presidency Project, UC Santa Barbara, November 5, 2021, https://www.presidency.ucsb.edu/documents/remarks-the-vice-president-during-visit-nasa-goddard-space-flight-center-greenbelt.

12. "Understanding the Multi-Trillion Dollar Space Economy—How Can the U.S. Successfully Compete With China," American Foreign Policy Council Capitol Hill Briefing, October 27, 2021, https://www.afpc.org/news/events/afpc-capitol-hill-briefing-understanding-the-multi-trillion-dollar-space-economyhow-can-the-u.s-successfully-compete-with-china.

13. Ibid.

14. See James Black, Linda Slapakova, and Kevin Martin, "Future Uses of Space Out to 2050: Emerging Threats and Opportunities for the UK National Space Strategy," RAND Corporation, 2022, https://www.rand.org/content/dam/rand/pubs/research_reports/RRA600/RRA609-1/RAND_RRA609-1.pdf; "Space: The Dawn of a New Age"; and "The Role of Space in Driving Sustainability, Security, and Development on Earth," McKinsey & Company and the World Economic Forum, 2022, https://www.mckinsey.com/industries/aerospace-and-defense/our-insights/the-role-of-space-in-driving-sustainability-security-and-development-on-earth.

15. Micah Maidenberg, "SpaceX Returns Private Astronauts to Earth After Three Days in Orbit," *Wall Street Journal*, September 19, 2021, https://www.wsj.com/articles/spacex-returns-private-astronauts-to-earth-after-three-days-in-orbit-11632006524; Jackie Wattles, Fernando Alphonso III, and Meg Wagner, "SpaceX Launches First All-Tourist Crew Into Orbit," CNN, September 15, 2021, https://www.cnn.com/business/live-news/spacex-inspiration4-launch-live-09-15-21/index.html.

16. "Blue Origin Safely Launches Four Commercial Astronauts to Space and Back," Blue Origin, December 22, 2021, https://www.blueorigin.com/news/first-human-flight-updates.

17. Robert Z. Pearlman, "Blue Origin Launches Michael Strahan and Crew of 5 on Record-Setting Suborbital Spaceflight," Space.com, December 11, 2021, https://www.space.com/blue-origin-michael-strahan-new-shepard-record-launch.

18. Micah Maidenberg, "William Shatner Goes to Space on Blue Origin's Second Human Flight," *Wall Street Journal*, October 13, 2021, https://www.wsj.com/articles/jeff-bezos-blue-origin-set-to-send-william-shatner-to-edge-of-space-11634117401.

19. Steve Gorman, "Billionaire Branson Soars to Space Aboard Virgin Galactic Flight," Reuters, July 12, 2021, https://www.reuters.com/lifestyle/science/virgin-galactics-branson-ready-space-launch-aboard-rocket-plane-2021-07-11/.

20. "8 Crew Members Wanted!" dearMoon, https://dearmoon.earth/; Michael Sheetz, "Bezos' Blue Origin Is Building More Rockets to Meet 'Robust Demand' for Space Tourism, CEO Says," CNBC, February 17, 2022, https://www.cnbc.com/2022/02/17/bezos-blue-origin-building-more-space-tourism-rockets-ceo.html; Mike Wall, "Virgin Galactic Sells 100 New Tickets for Space Tourist Launches," Space.com, November 10, 2021, https://www.space.com/virgin-galactic-sells-100-space-tourist-tickets.

21. "Space Tourism World Market Report," StrategyR, Global Industry Analysts Inc., https://www.strategyr.com/market-report-space-tourism-forecasts-global-industry-analysts-inc.asp.

22. "Future of Space Tourism: Lifting Off? Or Has Covid-19 Stunted Adoption?" UBS, July 20, 2021, https://www.ubs.com/global/en/investment-bank/in-focus/2021/space-tourism.html.

23. "Axiom Secures $130m in Additional Funding," Axiom Space, https://www.axiomspace.com/press-release/series-b.

24. "Human Spaceflight," Axiom Space, https://www.axiomspace.com/human-spaceflight.

25. Ibid.

26. "Voyager Class Station," Gateway Spaceport, December 10, 2021, https://gatewayspaceport.com/voyager-station/; Jeff Spry, "Company Plans to Start Building Private Voyager Space Station With Artificial Gravity in 2025," Space.com, February 25, 2021, https://www.space.com/orbital-assembly-voyager-space-station-artificial-gravity-2025.

27. "Voyager Station," Gateway Spaceport, https://voyagerstation.com/visit-voyager; Spry, "Company Plans to Start Building Private Voyager Space Station With Artificial Gravity in 2025."

28. Francesca Street, "World's First Space Hotel Scheduled to Open in 2027," CNN, March 4, 2021, https://www.cnn.com/travel/article/voyager-station-space-hotel-scn/index.html.

29. Peter Garretson, interview with Dennis Wingo, "33. Dennis Wingo: The Strategic Importance of the Moon," *Space Strategy*, podcast audio, February 11, 2022, https://anchor.fm/afpcspacepod/episodes/33--Dennis-Wingo-The-Strategic-Importance-of-the-Moon-e1e9iop.

30. "The Gateway," Gateway Spaceport, December 10, 2021, https://gatewayspaceport.com/the-gateway/; "What Is the Gateway Spaceport?" Gateway Spaceport, YouTube video, 2:11, https://www.youtube.com/watch?v=pbyPUzZ3Of4.

31. Michael Sheetz, "The Pentagon Wants to Use Private Rockets Like SpaceX's Starship to Deliver Cargo Around the World," CNBC, June 4, 2021, https://www.cnbc.com/2021/06/04/us-military-rocket-cargo-program-for-spacexs-starship-and-others.html.

32. Michael Sheetz, "Super Fast Travel Using Outer Space Could Be $20 Billion Market, Disrupting Airlines, UBS Predicts," CNBC, March 18, 2019, https://www.cnbc.com/2019/03/18/ubs-space-travel-and-space-tourism-a-23-billion-business-in-a-decade.html; "Space: The Dawn of a New Age," 79.

33. Sandra Erwin, "SpaceX Wins $102 Million Air Force Contract to Demonstrate Technologies for Point-to-Point Space Transportation," *SpaceNews*, January 20, 2022, https://spacenews.com/spacex-wins-102-million-air-force-contract-to-demonstrate-technologies-for-point-to-point-space-transportation/.

34. "In-Space Servicing, Assembly, and Manufacturing National Strategy," National Science and Technology Council, April 2022, https://www.whitehouse.gov/wp-content/uploads/2022/04/04-2022-ISAM-National-Strategy-Final.pdf.

35. "Successful Docking Paves the Way for Future On-Orbit and Life-Extension Services Through Robotics," Northrup Grumman, April 12, 2021, https://news.northropgrumman.com/news/releases/northrop-grumman-and-intelsat-make-history-with-docking-of-second-mission-extension-vehicle-to-extend-life-of-satellite.

36. "Space: The Dawn of a New Age."

37. "Proving Satellite Servicing," NASA's Exploration & In-Space Services, https://nexis.gsfc.nasa.gov/osam-1.html; Jennifer Harbaugh, "On-Orbit Servicing, Assembly, and Manufacturing 2 (OSAM-2)," NASA, October 23, 2019, https://www.nasa.gov/mission_pages/tdm/osam-2.html.

38. Michael Johnson, "Made in Space-Building a Better Optical Fiber," NASA, March 6, 2019, https://www.nasa.gov/mission_pages/station/research/news/b4h-3rd/eds-mis-building-better-optical-fiber/.

39. Nancy Kotary and Sara Cody, "Aboard NASA's Perseverance Rover, Moxie Creates Oxygen on Mars," *MIT News*, April 21, 2021, https://news.mit.edu/2021/aboard-nasa-perseverance-mars-rover-moxie-creates-oxygen-0421.

40. "DARPA Kicks off Program to Explore Space-Based Manufacturing," Defense Advanced Research Projects Agency, U.S. Department of Defense, March 23, 2022, https://www.darpa.mil/news-events/2022-03-23.

41. Randy Korotev, "The Chemical Composition of Lunar Soil," Some Meteorite Information, Washington University in St. Louis, https://sites.wustl.edu/meteoritesite/items/the-chemical-composition-of-lunar-soil/#CaAl.

42. "Artemis: NASA's Crewed Mission Delayed Till 2026," *WION News*, March 4, 2022, https://www.wionews.com/science/artemis-nasas-crewed-mission-delayed-till-2026-458714.

43. "Earth and Moon Space Transportation," PERMANENT, https://www.permanent.com/space-transportation-earth-moon.html.

44. "Space: The Dawn of a New Age," 73.

45. Donald K. Yeomans, "Why Study Asteroids," Jet Propulsion Laboratory, NASA, April 1988, https://ssd.jpl.nasa.gov/sb/why_asteroids.html.

46. Taylor Dinerman and Glenn Harlan Reynolds, "Trump Opens Outer Space for Business," *Wall Street Journal*, April 19, 2020, https://www.wsj.com/articles/trump-opens-outer-space-for-business-11587316780.

47. Larry M. Wortzel and Kate Selley, "Breaking China's Stranglehold on the U.S. Rare Earth Elements Supply Chain," *Defense Technology Program Brief* No. 22, American Foreign Policy Council, April 2021, https://www.afpc.org/uploads/documents/Defense_Technology_Briefing_-_Issue_22.pdf.

48. Wendy Whitman Cobb, "How SpaceX Lowered Costs and Reduced Barriers to Space," The Conversation, December 17, 2020, https://theconversation.com/how-spacex-lowered-costs-and-reduced-barriers-to-space-112586.

49. Sam Dinkin, "A Lunar Vision at $2,000/kg," *Space Review*, December 6, 2004, https://www.thespacereview.com/article/284/1.

50. "Orbital Construction: DARPA Pursues Plan for Robust Manufacturing in Space," Defense Advanced Research Projects Agency, U.S. Department of Defense, February 5, 2021, https://www.darpa.mil/news-events/2021-02-05.

51. "Space: The Dawn of a New Age," 72.

52. Steve Taranovich, "Helium-3 and Lunar Power for Earth Reactors," *EDN*, March 15, 2013, https://www.edn.com/helium-3-and-lunar-power-for-earth-reactors/.

53. "Space: The Dawn of a New Age."

54. Carlos de Castro, Margarita Mediavilla, Luis Javier Miguel, and Fernando Frechoso, "Global Solar Electric Potential: A Review of Their Technical and Sustainable Limits," *Renewable and Sustainable Energy Reviews* 28 (2013): 824–835, https://content.csbs.utah.edu/~mli/Economics%207004/Castro%20et%20al-Global%20Solar%20Electric%20Potential.pdf.

55. Don Petitt, "The Tyranny of the Rocket Equation," NASA, May 1, 2012, https://www.nasa.gov/mission_pages/station/expeditions/expedition30/tryanny.html.

56. Ian Sample, "Protect Solar System From Mining 'Gold Rush,' Say Scientists," *Guardian* (London), May 12, 2019, https://www.theguardian.com/science/2019/may/12/protect-solar-system-space-mining-gold-rush-say-scientists.

57. "Space: The Dawn of a New Age."

58. To see an example of an asteroid capture demonstration, watch https://www.youtube.com/watch?v=3pndYPsmasY.

59. "Optical Mining™ for Space Resource Harvesting," Trans Astronautica Corporation, https://transastra.com/our-technologies/#Optical-Mining.

60. "Omnivore Solar Thermal Rocket," Trans Astronautica Corporation, https://transastra.com/our-technologies/#Omnivore.

61. "Sutter Telescope Systems," Trans Astronautica Corporation, https://transastra.com/our-technologies/#Sutter-Telescope-Systems.

62. Joel Sercel, "Stepping Stones: Economic Analysis of Space Transportation Supplied From NEO Resources," Final Report on Grant No. NNX16AH11G Funded Under Economic Research for Space Development, NASA, October 15, 2017, https://www.nasa.gov/sites/default/files/atoms/files/eso_final_report.pdf.

63. John Brophy, Fred Culick, and Louis Friedman, *Asteroid Retrieval Feasibility Study* (Pasadena, CA: Jet Propulsion Laboratory, 2012).

64. The White House, "Executive Order: Encouraging International Support for the Recovery and Use of Space Resources," April 6, 2020, https://trumpwhitehouse.archives.gov/wp-content/uploads/2020/04/Fact-Sheet-on-EO-Encouraging-International-Support-for-the-Recovery-and-Use-of-Space-Resources.pdf.

65. Peter Garretson, interview with Joel Sercel, "'There Is Gold in Them Hills!': A Deep Dive Into Asteroid Mining & Space Logistics," *Space Strategy*, podcast audio, June 16, 2021, https://anchor.fm/afpcspacepod/episodes/There-is-Gold-in-Them-Hills--A-Deep-Dive-Into-Asteroid-Mining--Space-Logistics-e12tcfd.

66. "Space: The Dawn of a New Age," 67.

67. Rebecca Harrington, "This Incredible Fact Should Get You Psyched About Solar Power," *Business Insider*, September 29, 2015, https://www.businessinsider.com/this-is-the-potential-of-solar-power-2015-9.

68. Peter Garretson, "A Game Changer in the Fight Against Climate Change," *Newsweek*, April 5, 2022, https://www.newsweek.com/game-changer-fight-against-climate-change-opinion-1693929.

69. "The Role of Space in Driving Sustainability, Security, and Development on Earth," 25.

70. John Mankins, "SPS-Alpha: The First Practical Solar Power Satellite Via Arbitrarily Large Phased Array," NASA, September 15, 2020, https://www.nasa.gov/sites/default/files/atoms/files/niac_2011_phasei_mankins_spsalpha_tagged.pdf.

71. John Mankins, "SPS-ALPHA Mark-III and an Achievable Roadmap to Space Solar Power," John C. Mankins Click Space Technology.

72. Tim Hornyak, "Energy—Roping the Sun," *Scientific America*, July 1, 2008, https://www.scientificamerican.com/article/farming-solar-energy-in-space/.

73. Leonard David, "Space Solar Power's Time May Finally Be Coming," Space.com, November 3, 2021, https://www.space.com/space-solar-power-research-advances.

74. Mike Wall, "Relativity Space Unveils Fully Reusable, 3D-Printed Terran R Rocket," Space.com, June 8, 2021, https://www.space.com/relativity-space-reusable-terran-r-rocket.

75. C. E. Bloomquist, "Survey of Satellite Power Stations," Energy Research and Development Administration, September 1976, https://www.osti.gov/servlets/purl/7307148.

76. "Radiofrequency (RF) Radiation," American Cancer Society, https://www.cancer.org/cancer/cancer-causes/radiation-exposure/radiofrequency-radiation.html.

77. "Substances Listed in the Fourteenth Report on Carcinogens," National Toxicology Program, November 3, 2016, https://ntp.niehs.nih.gov/ntp/roc/content/listed_substances_508.pdf.

78. Cody Retherford, "The Promise of Space-Based Solar Power," *Space Policy Review* No. 1, American Foreign Policy Council, September 2022, https://www.afpc.org/uploads/documents/Space_Policy_Review_-_issue_1_-_9.21.2022.pdf.

79. Leonard David, "Space-Based Solar Power Getting Key Test Aboard US Military's Mysterious X-37B Space Plane," Space.com, April 8, 2021, https://www.space.com/x-37b-space-plane-solar-power-beaming.

80. "Space Power Beaming," Air Force Research Laboratory, https://afresearchlab.com/technology/space-power-beaming/; see also "ARACHNE," Air Force Research Laboratory, https://afresearchlab.com/technology/arachne/.

81. Stephen Chen, "China to Harvest Sun's Energy in Space and Beam It to Earth for Power by 2030," *South China Morning Post*, August 17, 2021, https://www.scmp.com/news/china/science/article/3145237/china-aims-use-space-based-solar-energy-station-harvest-suns; Gao Ji, Hou Xinbin, and Wang Li, "Solar Power Satellites Research in China," *Space Journal*, no. 16 (Winter 2010), https://spacejournal.ohio.edu/issue16/ji.html; Andrew Jones, "China Aims for Space-Based Solar Power Test in LEO in 2028, GEO in 2030," *SpaceNews*, June 8, 2022, https://spacenews.com/china-aims-for-space-based-solar-power-test-in-leo-in-2028-geo-in-2030/.

82. "A Public/Private COTS-Type Program to Develop Space Solar Power," National Space Society, February 2020, https://space.nss.org/wp-content/uploads/NSS-Position-Paper-COTS-Type-Space-Solar-Power-2020.pdf.

83. "Radioisotope Thermoelectric Generators (RTGs)," NASA, September 25, 2018, https://solarsystem.nasa.gov/missions/cassini/radioisotope-thermoelectric -generator/.

84. "Department of Energy ETEC Closure Project," ETEC, U.S. Department of Energy, https://www.etec.energy.gov/Operations/Major_Operations/SNAP_ Overview.php.

85. Aria Bendix, "Russia Plans to Launch a Nuclear-Powered Spacecraft That Can Travel From the Moon to Jupiter," *Business Insider*, May 25, 2021, https://www .businessinsider.com/russia-nuclear-powered-spacecraft-moon-venus-jupiter -2021-5.

86. Stephen Chen, "China's Nuclear Spaceships Will 'Mine Asteroids, Fly Tourists' by 2040," *South China Morning Post*, July 20, 2018, https://www.scmp.com /news/china/policies-politics/article/2120425/chinas-nuclear-spaceships-will -be-mining-asteroids; Anatoly Zak, "Russia Reveals a Formidable Nuclear-Powered Space Tug," Russian Space Web, August 25, 2021, https://www.russianspaceweb .com/tem.html; Anagha Srikanth, "Russia to Launch Nuclear-Powered Spaceship to the Moon, on to Venus, Then Jupiter," *The Hill*, May 26, 2021, https:// thehill.com/changing-america/enrichment/arts-culture/555560-russia-to -launch-nuclear-powered-spaceship-to-the/.

87. Pat McClure, "Overview of Current Analysis for DRACO System Safety Analysis Report (S-SAR)," DARPA Briefing Prepared for NASA, Slide 11, February 15, 2022.

88. Kelly Sands, "Fission System to Power Exploration on the Moon's Surface and Beyond," NASA, November 19, 2021, https://www.nasa.gov/feature /glenn/2021/fission-system-to-power-exploration-on-the-moon-s-surface-and -beyond; J. Harbaugh, "Fission Surface Power," NASA, May 6, 2021, https:// www.nasa.gov/mission_pages/tdm/fission-surface-power/index.html.

89. Harbaugh, "Fission Surface Power"; L. Hall, "Kilopower," NASA, December 12, 2017, https://www.nasa.gov/directorates/spacetech/kilopower.

90. Y. Xia, J. Li, R. Zhai, J. Wang, B. Lin, and Q. Zhou, "Application Prospect of Fission-Powered Spacecraft in Solar System Exploration Missions," *Space: Science & Technology*, February 26, 2021, https://spj.sciencemag.org/journals /space/2021/5245136/.

91. Nathan Greiner and Tabitha Dodson, "Demonstration Rocket for Agile Cislunar Operations (DRACO)," Defense Advanced Research Projects Agency, U.S. Department of Defense, https://www.darpa.mil/program/demonstration -rocket-for-agile-cislunar-operations; "DARPA Selects Performers for Phase 1 of Demonstration Rocket for Agile Cislunar Operations (DRACO) Program," Defense Advanced Research Projects Agency, U.S. Department of Defense, April 12, 2021, https://www.darpa.mil/news-events/2021-04-12.

92. Fraser Cain, "Earth to Mars in 100 Days: The Power of Nuclear Rockets," Phys.org, July 1, 2019, https://phys.org/news/2019-07-earth-mars-days-power -nuclear.html.

93. Christopher Helman, "Fueled by Billionaire Dollars, Nuclear Fusion Enters a New Age," *Forbes*, January 4, 2022, https://www.forbes.com/sites /christopherhelman/2022/01/02/fueled-by-billionaire-dollars-nuclear-fusion -enters-a-new-age/.

94. The White House, "Executive Order on Promoting Small Modular Reactors for National Defense and Space Exploration," January 12, 2021, https://trumpwhitehouse.archives.gov/presidential-actions/executive-order-promoting-small-modular-reactors-national-defense-space-exploration/.

95. Brian Wang, "Exponential Industrialization of Space Is More Important Than Combat Lasers and Hypersonic Fighters," Next Big Future, October 26, 2017, https://www.nextbigfuture.com/2017/10/exponential-industrialization-of-space-is-more-important-than-combat-lasers-and-hypersonic-fighters.html.

CHAPTER 2

1. A discussion of comprehensive national power (CNP) is available in Ray S. Cline, *World Power Assessment 1977: A Calculus of Strategic Drift* (Boulder, CO: Westview Press, 1977), ix, 206. Cline's theory of CNP is set forth in Ray S. Cline, *World Power Assessment: A Calculus of Strategic Drift* (Boulder, CO: Westview Press, 1975). Cline's basic formula was Pp = (C + E + M) × (S + W), where Pp = perceived power, C = critical mass = population + territory, E = economic capability, M = military capability, S = strategic purpose, and W = will to pursue national strategy. In China, there are a number of studies of CNP. For China's interpretation of CNP and what it means, see Wu Chunqiu (吴春秋), *Grand Strategy: A Chinese View* (大战略论) (Beijing: PLA Military Science Press, 1998), 55–70. Also see Song Ruiyu et al., *Theories on Measuring CNP* (测量理论综 国力) (Wuhan: Hubei Education Press, 1994), and Yan Xuetong et al., *Rise of China: Evaluation of International Environment* (中国的崛起：国际环境评价) (Tianjin: Tianjin People's Publishing House, 1996). CNP is also discussed in Hu Angang and Men Honghua, "The Rising of Modern China: Comprehensive National Power and Grand Strategy," originally published in *Strategy & Management*, no. 3 (2002), https://myweb.rollins.edu/tlairson/china/chigrandstrategy.pdf.

2. China National Space Administration (国家航天局), "2016 White Paper on China Space (2016中国的航天 白皮书)," December 27, 2016, http://www.cnsa.gov.cn/n6758824/n6758845/c6772477/content.html.

3. Dean Cheng, *How China Has Integrated Its Space Program Into Its Broader Foreign Policy* (Montgomery, AL: China Aerospace Studies Institute, 2021), 3.

4. Nadège Rolland, "A Concise Guide to the Belt and Road Initiative," National Bureau of Asian Research, April 11, 2019, https://www.nbr.org/publication/a-guide-to-the-belt-and-road-initiative/; Jacob J. Lew and Gary Roughead, "China's Belt and Road—Implications for the United States," *Independent Task Force Report* No. 79 (New York: Council on Foreign Relations, 2021), https://www.cfr.org/report/chinas-belt-and-road-implications-for-the-united-states/.

5. Guo Xiaoqiang (郭晓琼), *China Russia Economic Cooperation in the New Era: New Trends and New Problems* (新时代俄中经贸合作： 新i趋势与新问题) (Beijing: Chinese Academy of Social Sciences Press, 2021), 99, 151–152; see also Anastasiia Bondar, "China's Competitiveness in the Global Market: The OBOR Cooperation Program as a Tool for Its Implementation (中国在全球市场上的竞争力：以'一带一路'合作倡议为施行工具)," East China Normal University master's degree dissertation (华东师范大学硕士学位论文), May 20, 2020, CNKI.net.

6. Deng Liqun, ed., *China Today: Defense Science and Technology*, vol. 1 (Beijing: National Defense Industry Press, 1993), 356, cited in Cheng, *How China Has Integrated Its Space Program Into Its Broader Foreign Policy*, 22.

7. Ajey Lele, *China's 2016 Space White Paper: An Appraisal* (New Delhi: Institute for Defence Studies and Analysis, 2017), https://idsa.in/issuebrief/china-2016-space-white-paper_avlele_060117.

8. Mark Stokes and Dean Cheng, *China's Evolving Space Capabilities: Implications for U.S. Interests*, Report Prepared for the U.S.-China Economic and Security Review Commission (Alexandria, VA: Project 2049 Institute, April 26, 2012), 6, https://www.uscc.gov/sites/default/files/Research/USCC_China-Space-Program-Report_April-2012.pdf.

9. The State Council Information Office of the People's Republic of China, "Full Text: China's Space Program: A 2021 Perspective," January 28, 2022, http://english.www.gov.cn/archive/whitepaper/202201/28/content_WS61f35b3dc6d09c94e48a467a.html.

10. Larry M. Wortzel, "The Chinese People's Liberation Army and Space Warfare," *Astropolitics* 6, no. 2 (May–August 2008): 112. A PDF version of this article can be found at https://www.aei.org/research-products/working-paper/the-chinese-peoples-liberation-army-and-space-warfare/.

11. Larry M. Wortzel, "Cyber Espionage and the Theft of U.S. Intellectual Property and Technology," Testimony Before the House of Representatives on Energy and Commerce Subcommittee on Oversight and Investigations, July 9, 2013, https://www.uscc.gov/sites/default/files/Wortzel-OI-Cyber-Espionage-Intellectual-Property-Theft-2013-7-9.pdf; Larry M. Wortzel, "Export Controls on Satellite Technology," Testimony Before the House Committee on Foreign Affairs, Subcommittee on Terrorism, Nonproliferation and Trade, April 2, 2009, https://www.congress.gov/event/111th-congress/house-event/LC4897/text.

12. State Council of the People's Republic of China, "Full Text of White Paper on China's Space Activities in 2016," *China Daily*, December 28, 2016, http://english.www.gov.cn/archive/white_paper/2016/12/28/content_281475527159496.htm.

13. "China Advances Space Cooperation in 2020: Blue Book," *Xinhua*, March 10, 2021, http://www.china.org.cn/china/2021-03/10/content_77294512.htm.

14. Victor Tangermann, "China's New Rocket Concept Looks Suspiciously Like SpaceX's Starship," *Futurism*, February 18, 2022, https://futurism.com/the-byte/china-rocket-spacex-starship; Andrew Jones, "China Wants Its New Rocket for Astronaut Launches to Be Reusable," Space.com, March 6, 2022, https://www.space.com/china-reusable-rockets-for-astronaut-launches.

15. Irina Liu et al., "Commercial Space Policies and Drivers in China," *Evaluation of China's Commercial Space Sector*, Institute for Defense Analyses (2019), http://www.jstor.com/stable/resrep22872.5.

16. U.S. Department of State, "Military-Civil Fusion and the People's Republic of China," undated paper, https://www.state.gov/wp-content/uploads/2020/05/What-is-MCF-One-Pager.pdf; see also Larry M. Wortzel, "The Limitations of Military-Civil Mobilization: Problems With Funding and Clashing Interests in Industry-Based PLA Reserve Units," *China Brief* 19, no. 18 (October 18, 2019): 13–24; and Taylor A. Lee and Peter W. Singer, "China's Space Program Is More Military

Than You Might Think," *Defense One*, July 16, 2021, https://www.defenseone.com /ideas/2021/07/chinas-space-program-more-military-you-might-think/183790/.

17. Feng Songjiang (丰松江) and Dong Zhenghong (董正宏), *Space, The Future Battlefield!?: The New Situation and New Trends in the U.S. Militarization of Space* (太空，未来战场！？:美国太空军事化新态势新走向) (Beijing: Current Affairs Press, 2021), 138–139.

18. Long Kun (龙坤), Zhu Qichao (朱启超), Chen Xi (陈曦), and Ma Ning (马克宁), "The Trump Administration's Space Defense Strategy Adjustment: Motivations, Characteristics and Implications (特朗普政府太空防卫战略调整的动因特点与影响)," *National Defense Technology* (国防科技) 42, no. 4 (August 2021): 76–84.

19. Leng Mei (冷妹) and Ma Huijun (马惠军), "Studies on Motive Forces for Military-Civil Cooperation in the National Defense Industry (国防科技工业军民协同动力研究)," *National Defense Technology* (国防科技) 42, no. 4 (August 2021): 123–128.

20. Li Xueren (李学仁) et al., *Introduction to Military Aerospace and Space* (军事航空航天概论) (Xi'An: Northwest Industrial University Press, 2019), 222–228.

21. Information provided in this section was derived from the Union of Concerned Scientists Satellite Database, published on January 1, 2022, https://www .ucsusa.org/resources/satellite-database.

22. Larry Press, "A New Chinese Broadband Satellite Constellation," CircleID, October 2, 2020, https://circleid.com/posts/20201002-a-new-chinese-broadband -satellite-constellation/.

23. For a description of SBIRS, see "Space-Based Infrared Surveillance (SBIRS)," Lockheed Martin, https://www.lockheedmartin.com/en-us/products/sbirs.html.

24. See Larry M. Wortzel, "China's Nuclear Forces: Operations, Training, Doctrine, Command, Control and Campaign Planning," Strategic Studies Institute, U.S. Army War College, May 1, 2007, https://ssi.armywarcollege.edu/2007 /pubs/chinas-nuclear-forces-operations-training-doctrine-command-control -and-campaign-planning/; James R. Holmes, "China's Worrisome Edge Toward a 'Launch-on-Warning' Posture," *The Hill*, September 9, 2020, https://thehill.com /opinion/national-security/515243-chinas-worrisome-edge-toward-a-launch-on -warning-nuclear-posture.

25. Emmanuel Emmuet, "The Chinese Global Positioning System and the Convergence Between Electronic Warfare and Cyber Attack," *Asia Focus* No. 141 (Paris: Institute de Relations Internationales et Strategiques, May 2020), https://www.iris -france.org.

26. "China Promotes Mass Application for BeiDou Navigation System," *Global Times*, January 29, 2022, https://www.globaltimes.cn/page/202201/1250221.shtml.

27. John Xie, "China's Rival to GPS Navigation Carries Big Risks," *Voice of America News*, July 8, 2020, https://www.voanews.com/a/east-asia-pacific_voa-news -china_chinas-rival-gps-navigation-carries-big-risks/6192460.html.

28. Jordan Wilson, "China's Alternative to GPS and Its Implications for the United States," U.S.-China Economic and Security Review Commission, January 5, 2017, https://www.uscc.gov/research/chinas-alternative-gps-and-its-implications -united-states.

29. Evan Gough, "The Lunar Gateway Will Be in a 'Near-Rectilinear Orbit,'" *Universe Today*, July 19, 2019, https://www.universetoday.com/142896/the-Lunar -gateway-will-be-in-a-near-rectilinear-halo-orbit/.

30. Steffi Paladini, "China Could Gain a Monopoly on Space Stations. Here's What to Expect," *Space.com*, July 12, 2021, https://www.msn.com/en-us/news/technology/china-could-gain-a-monopoly-on-space-stations-heres-what-to-expect/ar-AAM3MTj.

31. Kenneth Chang, "China's Tiangong, First Space Station, Crashes Into Pacific," *New York Times*, April 1, 2018, https://www.nytimes.com/2018/04/01/science/chinese-space-station-crash-tiangong.html.

32. The Tiangong Space Station (中国空间站), Baidu Baike, https://baike.baidu.com/item/%E4%B8%AD%E5%9B%BD%E7%A9%BA%E9%97%B4%E7%AB%99/6287565.

33. Alberto Cervantes, Natasha Khan, and Danny Dougherty, "China's Space Station Tiangong Is Coming Together Bit by Bit," *Wall Street Journal*, September 17, 2021, https://www.wsj.com/articles/chinas-space-station-tiangong-is-coming-together-bit-by-bit-11631871002.

34. See the NASA catalog of orbits at https://earthobservatory.nasa.gov/features/OrbitsCatalog.

35. Cervantes et al., "China's Space Station Tiangong Is Coming Together Bit by Bit."

36. "China's Plans for the Moon, Mars and Beyond," *BBC News*, April 19, 2016, https://www.bbc.com/news/av/world-asia-36085659.

37. Andrew Jones, "China, Russia Enter MoU on International Lunar Research Station," *SpaceNews*, March 9, 2021, https://spacenews.com/china-russia-enter-mou-on-international-lunar-research-station/; Sputnik, "'Moon Base': Russia, China to Sign Pact on Lunar Research Station by End of 2022-Chinese Space Agency," *EurAsian Times*, April 8, 2022, https://eurasiantimes.com/russia-china-to-sign-pact-on-lunar-research-station-by-end-of-2022/; The State Council Information Office of the People's Republic of China, "Full Text: China's Space Program: A 2021 Perspective"; see also "International Lunar Research Station ILRS Guide for Partnership," Chinese National Space Administration, June 16, 2021, http://www.cnsa.gov.cn/english/n6465652/n6465653/c6812150/content.html.

38. Leonard David, "NASA Mulls Deep-Space Station on Moon's Far Side," Space.com, October 2, 2012, https://www.space.com/17856-nasa-deep-space-station-moon-farside.html.

39. Conor Skelding, "China's Mars Rover Completes 90-Day Mission," *New York Post*, August 21, 2021, https://nypost.com/2021/08/21/chinas-mars-rover-completes-90-day-mission/.

40. Matt Williams, "China Wants to Build a Spaceship That's Kilometers Long," Phys.org, September 1, 2021, https://phys.org/news/2021-09-china-spaceship-kilometers.html.

41. "Toward the 'Heavens One' Landing! 'Zhurong' Opened the Door to Mars, China Broke Through the Most Difficult Link (天问一号 登陆！"祝融"号叩开火星大门，中国突破最难的环节)," QQ.com, May 15, 2021, https://new.qq.com/omn/20210515/20210515A0306500.html.

42. "The Mars Rover Is on Temporary 'Vacation' Due to the Impact of the Sun and Is Expected to Resume Communications in Mid-October ("祝融号"火星车因日凌影响暂时"休假" 预计10月中旬将恢复通信)," *CCTV News*, October 9, 2021, https://news.cctv.com/2021/10/09/ARTIfbU9oSiZQ7rDNhWGEomN211009.shtml.

43. "China's Plans for the Moon, Mars and Beyond."

44. Jared Thompson, "Beijing's Troubling Space Ambitions," *SpaceNews*, May 20, 2021, https://spacenews.com/op-ed-beijings-troubling-space-ambitions/.

45. Lee Levkowitz and Nathan Beauchamp-Mustafaga, "China's Rare Earths Industry and Its Role in the International Market," U.S.-China Economic and Security Review Commission, November 3, 2010, https://www.uscc.gov/sites /default/files/Research/RareEarthsBackgrounderFINAL.pdf; Xie Kaifei (谢开飞), "What Makes the Rare Earth Trump Card (是什么成就了稀土这张王牌)," *Science and Technology Daily* (科技日报), June 6, 2019, http://www.xinhuanet.com /tech/2019-06/06/c_1124588710.htm.

46. Li Xiaoyu (李晓宇), "Study on Guidance and Control Methods for Asteroid Soft Landing (小行星软着陆动力下降段制导与控制方法研究)," master's thesis, Harbin University of Technology, 2015, https://xuewen.cnki.net/CMFD -1015980317.nh.html; Xiao Zhiyong (肖智勇), "Comparison Between Copernican-Aged Geological Activity on the Moon and Kuiperian-Aged Geological Activity on Mercury (月球表面哥白尼纪与水星表面柯伊伯纪的地质活动对比研究)," doctoral dissertation, China University of Geosciences, 2013, https://oversea-cnki-net .proxy.lib.ohio-state.edu/KCMS/detail/detail.aspx?dbcode=CDFD&dbnam e=CDFD1214&filename=1014150540.nh&uniplatform=NZKPT&amp-;v=v%25mmd2Fc6O3CyZ%25mmd2BY4DO0Ea6tX1LFB4nQQgMlWqIKTVgPKy SmmLByDxxHC7caiVdB63NO2.

47. "Mars, Here We Are! Space Mining May Become a Reality! (火星，我们来了！太空采矿或成为现实!)," *Mine Library Network*, 2020, https://zhuanlan.zhihu.com /p/136557257.

48. Jiao Yushu (焦玉书), "Landing on the Moon——Going to the Moon to Mine (登月——到月球去采矿)," *China Metallurgical News*, 2003, https://oversea-cnki -net.proxy.lib.ohio-state.edu/KCMS/detail/detail.aspx?dbcode=CCND&a mp;dbname=CCND0005&filename=CYJB20031029ZZ11&un iplatform=NZKPT&v=0Jq5dSF0eh%25mmd2FDohVaiwMw2VnGSK%- 25mmd2BYniPLANiQj6kNT90eV%25mmd2FOQZDoA2v5%25mmd2BSmv4fA5 DkSbA8%25mmd2BrBiCQ%3d.

49. Huang Haihua (黄海华), "What Elements and Minerals Are on the Surface of Mars? 'Hot and Golden Eyes' Completes a Detection in 300 Milliseconds (火星表面有哪些元素和矿物？"火眼金睛"300毫秒完成一次探测)," *Liberation Daily*, 2021, https://m.gmw.cn/baijia/2021-06/11/1302353531.html.

50. "China to Build Space-Based Solar Power Station by 2035," *China Daily*, December 2, 2019, https://www.chinadaily.com.cn/a/201912/02/WS5de47aa8a 310cf3e3557b515.html.

51. "China Reveals Plans to Launch a Fleet of Mile-Long Solar Panels Into Space," *Today's UK News*, August 18, 2021, https://todayuknews.com/science/china -reveals-plans-to-launch-a-fleet-of-mile-long-solar-panels-into-space/; "Exploiting Earth-Moon Space: China's Ambition After Space Station," *China Daily*, March 8, 2016, https://www.chinadaily.com.cn/china/2016-03/08/content_23775949 .htm.

52. Stephen Chen, "China Aims to Use Space-Based Solar Energy Station to Harvest Sun's Rays to Help Meet Power Needs," *South China Morning Post*, August 17, 2021, https://www.scmp.com/news/china/science/article/3145237/china -aims-use-space-based-solar-energy-station-harvest-suns; Gao Ji, Hou Xinbin, and

Wang Li, "Solar Power Satellites Research in China," *Online Journal of Space Communication* 9, no. 16 (Winter 2010), https://ohioopen.library.ohio.edu/cgi/viewcontent.cgi?article=1398&context=spacejournal.

53. Ray Kwong, "China Is Winning the Solar Space Race," *Foreign Policy*, June 16, 2019, https://foreignpolicy.com/2019/06/16/china-is-winning-the-solar-space-race/.

54. Yanpei Tian, Xiaojie Liu, Fangqi Chen, and Yi Zheng, "Harvesting Energy From Sun, Outer Space, and Soil," *Scientific Reports* 10 (2020), article 20903, https://www.nature.com/articles/s41598-020-77900-7.

55. Stephen Chen, "China's Nuclear Spaceships Will Be 'Mining Asteroids and Flying Tourists' as It Aims to Overtake U.S. in Space Race," *South China Morning Post*, November 17, 2017, https://www.scmp.com/news/china/policies-politics/article/2120425/chinas-nuclear-spaceships-will-be-mining-asteroids.

56. See the testimony of Namrata Goswami before the U.S.-China Economic and Security Review Commission hearing on "China in Space: A Strategic Competition?" April 25, 2019, https://www.uscc.gov/sites/default/files/Namrata%20Goswami%20USCC%2025%20April.pdf.

57. Bart Hendrickx, "Ekipuzhi: Russia's Top-Secret Nuclear-Powered Satellite," *Space Review*, October 7, 2019, https://thespacereview.com/article/3809/1.

58. See Mark Stokes and Dean Cheng, *China's Evolving Space Capabilities: Implications for U.S. Interests*, A Report for the U.S.-China Economic and Security Review Commission (Arlington, VA: Project 2049 Institute, 2011), 13–17.

59. U.S. House of Representatives, *Report of the Select Committee on U.S. National Security and Military/Commercial Concerns With the People's Republic of China*, Submitted by Mr. Cox of California, Chairman (Washington, DC: U.S. Government Printing Office, 1999), https://www.congress.gov/105/crpt/hrpt851/CRPT-105hrpt851.pdf.

60. "China's Satellite Launch and Tracking Control General (CTLC)," NTI.org, January 31, 2013, https://www.nti.org/learn/facilities/124/. See also Mark A. Stokes, "The People's Liberation Army and China's Space and Missile Development: Lessons From the Past and Prospects for the Future," in Laurie Burkitt, Andrew Scobell, and Larry M. Wortzel, eds., *The Lessons of History: The Chinese People's Liberation Army at 75* (Carlisle, PA: Strategic Studies Institute, July 2003).

61. John Costello, "The Strategic Support Force: Update and Overview," *China Brief* 16, no. 19 (December 21, 2016), https://jamestown.org/program/strategic-support-force-update-overview/.

62. Ibid.

63. Feng Songjiang (丰松江) and Dong Zhenghong, *Space, The Future Battlefield!?: The New Situation and New Trends in the U.S. Militarization of Space* (董正宏) (太空，未来战场！？:美国太空军事化新态势新走向) (Beijing: Current Affairs Press, 2021).

64. Ibid., 15.

65. Larry M. Wortzel, "The Chinese People's Liberation Army and Space Warfare: Emerging United States-China Military Competition," American Enterprise Institute Working Paper, October 17, 2007, https://www.aei.org/research-products/working-paper/the-chinese-peoples-liberation-army-and-space-warfare/; and Craig Murray, "China Missile Launch May Have Tested Part of a New Anti-Satellite Capability," *U.S.-China Economic and Security Review Commission Staff Research Backgrounder*, May 22, 2013, https://www.uscc.gov/sites/default/files

/Research/China%20Missile%20Launch%20May%20Have%20Tested%20
Part%20of%20a%20New%20Anti-Satellite%20Capability_05.22.13.pdf.

66. Mandy Mayfield, "China's Ambitious Space Programs Raise Red Flags," *National Defense*, July 2, 2021 https://www.nationaldefensemagazine.org/articles /2021/7/2/chinas-ambitious-space-programs-raise-red-flags.

67. Demetri Sevastopulo, "China Conducted Two Hypersonic Weapons Tests This Summer," *Financial Times*, October 20, 2021, https://www.ft.com/content /c7139a23-1271-43ae-975b-9b632330130b.

68. Demetri Sevastopulo and Kathrin Hille, "China Tests New Space Capability With Hypersonic Missile," *Financial Times*, October 16, 2021, https://www.ft.com /content/ba0a3cde-719b-4040-93cb-a486e1f843fb.

69. "Beijing Mocks America Saying Their New 21,000 mph Missile Is a 'New Blow to the US,'" *Daily Mail (UK)*, October 18, 2021, https://www.dailymail .co.uk/news/article-10101953/Beijing-mocks-America-saying-new-21-000mph -nuclear-capable-missile-new-blow-US.html.

70. Sevastopulo and Hille, "China Tests New Space Capability With Hypersonic Missile."

71. Mathilde Minet, "Understanding the Wolf Agreement," *Space Legal Issues*, October 25, 2020, https://www.spacelegalissues.com/understanding-the-wolf -agreement/.

72. Cheng, *How China Has Integrated Its Space Program Into Its Broader Foreign Policy*, 16; "The Belt and Road Initiative: Progress, Contributions and Prospects," Permanent Mission of the People's Republic of China to the United Nations Office at Geneva and Other International Organizations in Switzerland, https://www .mfa.gov.cn/ce/cegv/eng/zywjyjh/t1675564.htm.

73. Jevans Nyabiage, "China Boosts Its Soft Power in Africa While Launching African Space Ambitions," *South China Morning Post*, October 11, 2020, https://www.scmp.com/news/china/diplomacy/article/3104900/china-boosts -its-soft-power-africa-while-launching-african.

74. Julie Michelle Klinger, "China and Africa in Global Space Science, Technology, and Satellite Development," Working Paper No. 45 (Washington, DC: School of Advanced International Studies, Johns Hopkins University, 2020), https://static1. squarespace.com/static/5652847de4b033f56d2bdc29/t/5ecdba4495844040ae1 a6c60/1590540868709/PB+45+-+Klinger+-+China+Africa+Space+Satellites.pdf.

75. "Swakopmund, Namibia," Global Security, https://www.globalsecurity. org/space/world/china/swakopmund.htm.

76. Nyabiage, "China Boosts Its Soft Power in Africa While Launching African Space Ambitions."

77. Ibid.

78. See U.S. Congress, Office of Technology Assessment, "History and Current Status of the Russian Space Program," in *U.S.-Russian Cooperation in Space*, OTA-ISS-618 (Washington, DC: U.S. Government Printing Office, April 1995), 26–36, 137–139, https://www.princeton.edu/~ota/disk1/1995/9546/954605.PDF; "A Guide to the 'Stans' of Central Asia," August 15, 2004, https://reconsideringrus-sia.org/2014/08/15/a-guide-to-the-stans-of-central-asia/.

79. Preethi Amaresh, "All Weather Friends: China and Pakistan Space Cooperation," *The Diplomat*, January 30, 2020, https://thediplomat.com/2020/01/ all-weather-friends-china-and-pakistan-space-cooperation/.

80. Eugene Rumer, Richard Sokolski, and Paul Stronski, *U.S. Policy Toward Central Asia 3.0* (Washington, DC: Carnegie Endowment for International Peace, 2016), 8–9.

81. Arjun Kharpal, "In Battle With U.S., China to Focus on 7 'Frontier' Technologies From Chips to Brain-Computer Fusion," CNBC, March 5, 2021, https://www.cnbc.com/2021/03/05/china-to-focus-on-frontier-tech-from-chips-to-quantum-computing.html.

82. Center for Security and Emerging Technology, Georgetown University, "Translation: Outline of the People's Republic of China 14th Five-Year Plan for National Economic and Social Development and Long-Range Objectives for 2035," National People's Congress of China, May 13, 2021, https://cset.georgetown.edu/publication/china-14th-five-year-plan/.

CHAPTER 3

1. For more information on the EMP threat, see Richard M. Harrison et al., "Strategic Primer: Electromagnetic Threats," American Foreign Policy Council 4 (Winter 2018), https://www.afpc.org/uploads/documents/EMP%20Primer%20-final.pdf.

2. "SpaceX Loses 40 Satellites to Geomagnetic Storm a Day After Launch," *BBC News*, February 9, 2022, https://www.bbc.com/news/world-60317806.

3. Jonathan O'Callaghan, "New Studies Warn of Cataclysmic Solar Superstorms," *Scientific American*, September 24, 2019, https://www.scientificamerican.com/article/new-studies-warn-of-cataclysmic-solar-superstorms/.

4. "A Near-Earth Asteroid Census," https://www.nasa.gov/mission_pages/WISE/multimedia/gallery/neowise/pia14734.html; Jordan Riley et al., "Directed Energy Active Illumination for Near-Earth Object Detection," https://digitalcommons.calpoly.edu/cgi/viewcontent.cgi?referer=&httpsredir=1&article=1044&context=stat_fac.

5. Elizabeth Howell, "Chelyabinsk Meteor: A Wake-Up Call for Earth," Space.com, January 9, 2019, https://www.space.com/33623-chelyabinsk-meteor-wake-up-call-for-earth.htm.

6. *National Aeronautics and Space Administration Authorization Act of 2008*, H.R. 6063, 110th Congress (2007–2008), https://www.congress.gov/bill/110th-congress/house-bill/6063.

7. Stephen M. McCall, "Challenges to the United States in Space," Congressional Research Service Report, January 27, 2020, https://crsreports.congress.gov/product/pdf/IF/IF10337.

8. For information on space sustainability challenges, see Secure World Foundation, "Space Sustainability Challenges," https://swfound.org/resource-library/space-sustainability-challenges/; George E. Pollock IV and James A. Vedda, "Cislunar Stewardship: Planning for Sustainability and International Cooperation," Center for Space Policy and Strategy, Aerospace Corporation, June 2020, https://aerospace.org/sites/default/files/2020-06/Pollock-Vedda_CislunarStewardship_20200601.pdf; and for information on space situational awareness and space traffic management, see https://swfound.org/resource-library/space-situational-awareness-and-space-traffic-management/.

9. There are several expert threat assessments that outline potential U.S. adversary space doctrine. See, for example, Brian Weeden and Victoria Samson, "Global Counterspace Capabilities: An Open Source Assessment," Secure World Foundation, April 2020, https://swfound.org/counterspace; Todd Harrison, Kaitlyn Johnson, Thomas G. Roberts, Tyler Way, and Makena Young, "Space Threat Assessment 2020," Center for Strategic and International Studies, March 2020, https://www.csis.org/analysis/space-threat-assessment-2020; "Challenges to Security in Space," U.S. Defense Intelligence Agency, January 2019, https://media.defense.gov/2019/Feb/11/2002088710/-1/-1/1/SPACE-SECURITY-CHALLENGES.PDF; Zachary Wilson, "Competing in Space," National Air and Space Intelligence Center, January 16, 2019, https://media.defense.gov/2019/Jan/16/2002080386/-1/-1/1/190115-F-NV711-0002.PDF; and Elbridge Colby, "From Sanctuary to Battlefield: A Framework for a U.S. Defense and Deterrence Strategy for Space," Center for a New American Security, January 2016, https://www.cnas.org/publications/reports/from-sanctuary-to-battlefield-a-framework-for-a-us-defense-and-deterrence-strategy-for-space.

10. "Cyberspace Operations," Joint Publication 3-12, June 8, 2018, https://www.jcs.mil/Portals/36/Documents/Doctrine/pubs/jp3_12.pdf.

11. "Challenges to Security in Space."

12. Ibid.

13. Ibid.; G. J. Nunz, "Beam Experiments Aboard a Rocket (BEAR) Project Final Report, Vol. 1: Project Summary," Prepared for the Strategic Defense Initiative Organization Directed Energy Office, January 1, 1990, https://apps.dtic.mil/dtic/tr/fulltext/u2/a338597.pdf.

14. See Harrison et al., "Strategic Primer: Electromagnetic Threats."

15. "Challenges to Security in Space."

16. Stephen Clark, "Upgraded X-37B Spaceplane Rockets Into Orbit Aboard Atlas 5 Launcher," *Spaceflight Now*, May 17, 2020, https://spaceflightnow.com/2020/05/17/upgraded-x-37b-spaceplane-rockets-into-orbit-aboard-atlas-5-launcher/.

17. For more information on the threat from hypersonic weapons, see Richard M. Harrison et al., "Strategic Primer: Hypersonic Weapons," American Foreign Policy Council 6 (Summer 2019), https://www.afpc.org/uploads/documents/Hypersonic_Weapons_Primer_-_July_2019_(web).pdf.

18. "2022 Challenges to Security in Space," U.S. Defense Intelligence Agency, March 2022, https://www.dia.mil/Portals/110/Documents/News/Military_Power_Publications/Challenges_Security_Space_2022.pdf.

19. Abigail Beall, "China's Private Space Industry Is Rapidly Gaining Ground on SpaceX," Wired (UK), August 28, 2019, https://www.wired.co.uk/article/china-private-space-industry.

20. "Company Profile," China Aerospace Science and Technology Corporation, 2018, http://english.spacechina.com/n16421/n17138/n17229/index.html; "Introduction of CASIC," China Aerospace Science and Industry Corporation Ltd., 2018, http://www.casic.com/n189298/n189314/index.html.

21. Josh Baughman, "China's Satellite Super Factories and US National Security," *Cyber: The Magazine of the MCPA*, July 22, 2021, https://public.milcyber.org/activities/magazine/articles/2021/baughman-chinas-satellite-super-factories.

22. "2022 Challenges to Security in Space."

23. "Challenges to Security in Space."

24. Andrew Jones, "China Closes Record-Breaking Year With Orbital Launches From Jiuquan and Xichang," *SpaceNews*, December 30, 2021, https://spacenews.com/china-closes-record-breaking-year-with-orbital-launches-from-jiuquan-and-xichang/.

25. Andrew Jones, "China Outlines Intense Space Station Launch Schedule, New Astronaut Selection," *SpaceNews*, May 28, 2020, https://spacenews.com/china-outlines-intense-space-station-launch-schedule-new-astronaut-selection/.

26. Jessie Yeung, "China and Russia Agree to Build Joint Lunar Space Station," *CNN*, March 10, 2021, https://www.cnn.com/2021/03/09/asia/russia-china-lunar-station-intl-hnk-scli-scn/index.html.

27. "Challenges to Security in Space."

28. Ben Turner, "Chinese Scientists Call for Plan to Destroy Elon Musk's Starlink Satellites," *Live Science*, May 27, 2022, https://www.livescience.com/china-plans-ways-destroy-starlink.

29. Andrew Jones, "China Carries Out Secretive Launch of 'Reusable Experimental Spacecraft,'" *SpaceNews*, September 4, 2020, https://spacenews.com/china-carries-out-secretive-launch-of-reusable-experimental-spacecraft/.

30. Demetri Sevastopulo and Kathrin Hille, "China Tests New Space Capability With Hypersonic Missile," *Financial Times*, October 16, 2021, https://www.ft.com/content/ba0a3cde-719b-4040-93cb-a486e1f843fb.

31. Andrew Jones, "China's CASIC Reveals Five-Year Plan for Reusable Spaceplane, Commercial Space Projects," *SpaceNews*, October 19, 2020, https://spacenews.com/chinas-casic-reveals-five-year-plan-for-reusable-space-plane-commercial-space-projects/.

32. Jones, "China Carries Out Secretive Launch of 'Reusable Experimental Spacecraft.'"

33. For more information on Chinese companies transferring technology, see Sean O'Connor, "How Chinese Companies Facilitate Technology Transfer From the United States," U.S.-China Economic and Security Review Commission, Staff Research Report, May 6, 2019, https://www.uscc.gov/sites/default/files/Research/How%20Chinese%20Companies%20Facilitate%20Tech%20Transfer%20from%20the%20US.pdf.

34. For more information on the CCP Three Warfares strategy, see Peter Mattis, "China's 'Three Warfares' in Perspective," *War on the Rocks*, January 13, 2018, https://warontherocks.com/2018/01/chinas-three-warfares-perspective/.

35. "UK and US Say Russia Fired a Satellite Weapon in Space," *BBC News*, July 23, 2020, https://www.bbc.com/news/world-europe-53518238; U.S. Department of State, "Russia Conducts Destructive Anti-Satellite Missile Test," Press Release by Secretary of State Antony J. Blinken, November 15, 2021, https://www.state.gov/russia-conducts-destructive-anti-satellite-missile-test/.

36. Anatoly Zak, "Russia Approves Its 10-Year Space Strategy," Planetary Society, March 23, 2016, https://www.planetary.org/articles/0323-russia-space-budget.

37. "Reports," BryceTech, 2021, https://brycetech.com/reports.

38. "Challenges to Security in Space."

39. Meghan Bartels, "Russia Is Going Back to the Moon This Year," Space.com, April 15, 2021, https://www.space.com/russia-luna-25-returning-to-moon;

Tanmay Kadam, "Lunar Research Station: Russia, China Almost Ready to Ink Pact on 'Moon Base' That Will Rival Artemis Accords–Rogozin," *EurAsian Times*, https://eurasiantimes.com/russia-china-almost-ready-to-ink-pact-on-moon-base/.

40. Tom Balmforth, "Russia Has Begun Spaceplane Project, Says Soviet Shuttle Designer," Reuters, March 24, 2021, https://www.reuters.com/article/us-space-russia/russia-has-begun-spaceplane-project-says-soviet-shuttle-designer-idUSKBN2BG1YR.

41. Patrick Tucker, "Russia Might Try Reckless Cyber Attacks as Ukraine War Drags On, US Warns," *Defense One*, June 15, 2021, https://www.defenseone.com/threats/2022/06/russia-might-try-reckless-cyber-attacks-ukraine-war-drags-us-warns/368242/; Jason Rainbo, "As U.S. Blames Russia for KA-SAT Hack, Starlink Sees Growing Threat," *SpaceNews*, May 11, 2022, https://spacenews.com/as-us-blames-russia-for-ka-sat-hack-starlink-sees-growing-threat/.

42. Hanneke Weitering, "Russia Has Launched an Anti-Satellite Missile Test, US Space Command Says," Space.com, December 16, 2020, https://www.space.com/russia-launches-anti-satellite-missile-test-2020.

43. "Worldwide Threat Assessment of the US Intelligence Community," Office of the Director of National Intelligence, February 2018, https://www.dni.gov/files/documents/Newsroom/Testimonies/2018-ATA---Unclassified-SSCI.pdf.

44. Tyler Rogoway and Ivan Voukadinov, "Exclusive: Russian MiG-31 Foxhound Carrying Huge Mystery Missile Emerges Near Moscow," The Drive, September 29, 2018, https://www.thedrive.com/the-war-zone/23936/exclusive-russian-mig-31-foxhound-carrying-huge-mystery-missile-emerges-near-moscow.

45. Yuri Karash, "Russian Space Program: Financial State, Current Plans, Ambitions and Cooperation With the United States," *Space Congress Proceedings* 27 (2016), https://commons.erau.edu/space-congress-proceedings/proceedings-2016-44th/presentations-2016/27.

46. "В Роскосмосе Сравнили Свой БЮДЖЕТ и NASA," TASS, February 11, 2020, https://tass.ru/ekonomika/7734535.

47. Zak, "Russia Approves Its 10-Year Space Strategy."

48. Pavel Luzin, "Roscosmos Suffers From Russia's Confrontation With the US," Eurasia Daily Monitor, June 22, 2021, https://jamestown.org/program/roscosmos-suffers-from-russias-confrontation-with-the-us/.

49. Colin Zwirko, "North Korea Holds Space Conference, Says Launching Satellites Will Help Economy," *NK News*, November 22, 2021, https://www.nknews.org/2021/11/north-korea-holds-space-conference-says-launching-satellites-will-help-economy/.

50. Robert Z. Pearlman, "North Korea's 'Nada' Space Agency, Logo Are Anything but 'Nothing,'" Space.com, April 2, 2014, https://www.space.com/25337-north-korea-nada-space-agency-logo.html.

51. Geoff Brumfiel, "North Korea Seen Expanding Rocket Launch Facility It Once Promised to Dismantle," NPR National Security, March 27, 2020, https://www.npr.org/2020/03/27/822661018/north-korea-seen-expanding-rocket-launch-facility-it-once-promised-to-dismantle.

52. "North Korea Administrative Designations Updates," U.S. Department of the Treasury, May 13, 2020, https://home.treasury.gov/policy-issues/financial

-sanctions/recent-actions/20200513; "The United States Sanctions North Korean Government Officials and Organizations Tied to Its Missile and Nuclear Programs," U.S. Department of the Treasury, March 2, 2016, https://home.treasury.gov/news/press-releases/jl0372.

53. "Status of North Korean Satellite Unknown After Prolonged Radio Silence, Reports of Tumbling," Spaceflight101, February 12, 2016, https://spaceflight101.com/status-of-north-korean-satellite-unknown-after-prolonged-radio-silence-reports-of-tumbling/.

54. Ibid.; "North Korean Satellite 'Tumbling in Orbit,' U.S. Officials Say," *CBS News*, February 8, 2016, https://www.cbsnews.com/news/north-korea-satellite-tumbling-in-orbit-u-s-officials-say/.

55. Some experts, including Peter Pry, have disputed this and argued that the North Korean satellites could be electromagnetic-pulse weapons.

56. Hyung-Jin Kim, "N Korea Confirms Missile Tests as Biden Warns of Response," Associated Press, March 26, 2021, https://apnews.com/article/joe-biden-south-korea-north-korea-united-nations-pyongyang-4ff07ea48279a6d8d739415d-2bab9f06.

57. "Challenges to Security in Space."; "North Korean Tactics," *Army Technique Publications*, July 24, 2020, https://fas.org/irp/doddir/army/atp7-100-2.pdf.

58. "Iranian Space Agency," IAF, 2013, https://www.iafastro.org/membership/all-members/iranian-space-agency.html.

59. "Iran Space Research Center," Iran Watch, October 31, 2019, https://www.iranwatch.org/iranian-entities/iran-space-research-center.

60. "Iran-Related Designations; Non-Proliferation Designations; Kingpin Act Designations Update," U.S. Department of the Treasury, September 3, 2019, https://home.treasury.gov/policy-issues/financial-sanctions/recent-actions/20190903_33.

61. David Todd, "Iran Launches New Qased Rocket Type With Noor Military Sat Aboard," Seradata Space Intelligence, April 22, 2020, https://www.seradata.com/iran-launches-new-qased-rocket-type-with-noor-military-sat-aboard/.

62. "U.S. Condemns Iran's Satellite Launch," Iran Primer, April 27, 2020, https://iranprimer.usip.org/blog/2020/apr/27/us-condemns-iran's-satellite-launch.

63. Radio Farda, "Space Force Commander Says Iran's Military Satellite Launches Will Continue," Radio Farda, April 23, 2020, https://en.radiofarda.com/a/space-force-commander-says-iran-s-military-satellite-launches-will-continue/30573116.html.

64. Uzi Rubin, "Iran's Revolutionary Guard Goes to Space," Jerusalem Institute for Strategy and Security, April 29, 2020, https://jiss.org.il/en/rubin-irans-revolutionary-guard-goes-to-space/; Max Fisher, "Deep in the Desert, Iran Quietly Advances Missile Technology," *New York Times*, May 23, 2018, https://www.nytimes.com/2018/05/23/world/middleeast/iran-missiles.html.

65. Andrew Hanna, "Iran's Ambitious Space Program," Iran Primer, June 23, 2021, https://iranprimer.usip.org/blog/2020/jun/23/iran%E2%80%99s-ambitious-space-program.

66. "Iran Utilizes Satellite 'Noor' to Track Oil Tankers to Venezuela: IRGC Cmdr.," Mehr News Agency, June 8, 2020, https://en.mehrnews.com/news/159574/Iran-utilizes-satellite-Noor-to-track-oil-tankers-to-Venezuela.

67. "Iran's First Military Satellite Beams Images of US Bases in Middle East," DefenseWorld.net, July 31, 2020, https://www.usadefensenews.com/2020/07/31/irans-first-military-satellite-beams-images-of-us-military-bases-in-middle-east/.

68. J. D. Simkins, "Space Force General Trolls Iranian Military Satellite Launch—'Space Is Hard,'" *Military Times*, April 28, 2020, https://www.militarytimes.com/off-duty/military-culture/2020/04/28/space-force-general-trolls-iranian-military-satellite-launch-space-is-hard/.

69. "Challenges to Security in Space"; Nathan Strout, "Here's What You Need to Know About Iran's Counter-Space Weapons," C4ISRNet, January 8, 2020, https://www.c4isrnet.com/battlefield-tech/space/2020/01/08/what-we-know-about-irans-counter-space-weapons/.

70. Seth J. Frantzman, "Iran Claims to Pioneer New Electronic Warfare Unit," *Jerusalem Post*, July 7, 2019, https://www.jpost.com/middle-east/iran-claims-to-pioneer-new-electronic-warfare-unit-594858; "Iran Unveils IRGC Navy 'Missile City' Arsenal," *Jewish News Syndicate*, March 17, 2021, https://www.jns.org/iran-unveils-irgc-navy-missile-city-arsenal/.

71. Strout, "Here's What You Need to Know About Iran's Counter-Space Weapons."

72. M. J. Holzinger, C. C. Chow, and P. Garretson, "A Primer on Cislunar Space," Air Force Research Laboratory, May 3, 2021, https://www.afrl.af.mil/Portals/90/Documents/RV/A%20Primer%20on%20Cislunar%20Space_Dist%20A_PA2021-1271.pdf.

73. Spencer Kaplan, "Eyes on the Prize: The Strategic Implications of Cislunar Space and the Moon," Aerospace Security Project, Center for Strategic and International Studies, July, 13, 2020, http://aerospace.csis.org/wp-content/uploads/2020/07/20200714_Kaplan_Cislunar_FINAL.pdf.

74. "Chang'e-2 Moon Orbiter Reaches L2 Point," Solar System Exploration Research Virtual Institute, NASA, https://sservi.nasa.gov/articles/change-2-moon-orbiter-reaches-l2-point/.

75. Kaplan, "Eyes on the Prize."

76. Peter Garretson, interview with Bill Bruner, "Bull Bruner: Our Hypersonic Future," *Space Strategy*, podcast audio, July 22, 2021, https://anchor.fm/afpcspacepod/episodes/Bill-Bruner-Our-Hypersonic-Future-e14roq9.

77. Holzinger et al., "A Primer on Cislunar Space."

78. Debra Werner, "Updated Intelligence Report Calls for Improved Monitoring of Cislunar Space," *SpaceNews*, August 24, 2021, https://spacenews.com/dia-report-2021-cislunar-monitoring/.

79. Kaplan, "Eyes on the Prize."

80. Ibid.

81. Clementine G. Starling, Mark J. Massa, Christopher P. Mulder, and Julia T. Seigel, "The Future of Security in Space: A Thirty-Year US Strategy," Atlantic Council Strategy Paper, April 2021, https://www.atlanticcouncil.org/wp-content/uploads/2021/04/TheFutureofSecurityinSpace.pdf.

CHAPTER 4

1. Steve Garber, "Sputnik and the Dawn of the Space Age," NASA History, October 10, 2007, https://history.nasa.gov/sputnik/.

2. Alan Wasser, "LBJ's Space Race: What We Didn't Know Then (Part 1)," *Space Review*, June 20, 2005, https://www.thespacereview.com/article/396/1.

3. "Expert Sees Moon as Rocket Base," *Washington Post*, January 29, 1958, A1, https://search.proquest.com/docview/149092400/CEB96DEB0519483CPQ/1?ac countid=201395.

4. "Apollo 17 Lunar Module /ALSEP," NASA Space Science Data Coordinated Archive, https://nssdc.gsfc.nasa.gov/nmc/spacecraft/display.action?id =1972-096C.

5. Dennis Wingo, "The Early Space Age, the Path Not Taken Then, but Now? (Part I)," February 16, 2015, https://denniswingo.wordpress.com/2015/02/16 /the-early-space-age-the-path-not-taken-then-but-now/.

6. Amy Thompson, "Liftoff! SpaceX Launches 1st Astronauts for NASA on Historic Test Flight," Space.com, May 30, 2020, https://www.space.com /spacex-demo2-nasa-astronaut-launch-success.html.

7. Lara Seligman, "The New Space Race," *Foreign Policy*, May 14, 2019, https:// foreignpolicy.com/2019/05/14/the-new-space-race-china-russia-nasa/.

8. "NASA's Proud Space Shuttle Program Ends With Atlantis Landing," NASA, July 22, 2011, https://www.nasa.gov/home/hqnews/2011/jul/HQ_11-240 _Atlantis_Lands.html.

9. Dave Mosher and Hilary Brueck, "Astronauts Explain Why Nobody Has Visited the Moon in More Than 45 Years—and the Reasons Are Depressing," *Business Insider*, July 19, 2019, https://www.businessinsider.com/Moon-missions-why -astronauts-have-not-returned-2018-7; Matthew Shindell, "Waning Interest," *Distillations*, Science History Institute, July 19, 2016, https://www.sciencehistory.org /distillations/magazine/waning-interest; Clelia Iacomino and Silvia Ciccarelli, "Potential Contributions of Commercial Actors to Space Exploration," *Advances in Astronautics Science and Technology* 1 (August 24, 2018): 141–151, https://doi .org/10.1007/s42423-018-0011-7; Nicholas Eftimiades, "Small Satellites: The Implications for National Security," Atlantic Council, May 2022, https://www .atlanticcouncil.org/wp-content/uploads/2022/05/Small_satellites-Implications _for_national_security.pdf.

10. Courtney Johnson, "How Americans See the Future of Space Exploration, 50 Years After the First Moon Landing," Fact Tank News in the Numbers, Pew Research Center, July 17, 2019, https://www.pewresearch.org/fact-tank/2019/07/17/how -americans-see-the-future-of-space-exploration-50-years-after-the-first-Moon -landing/.

11. Anthony Imperato, Peter Garretson, and Richard M. Harrison, "U.S. Space Budget Report," *Defense Technology Program Brief* No. 23, American Foreign Policy Council, May 2021, https://www.afpc.org/uploads/documents/Defense _Technology_Briefing_-_Issue_23.pdf.

12. Anthony Imperato, Peter Garretson, and Richard M. Harrison, "To Compete With China in Space, America Must Ramp Up Funding," *National Interest*, June 1, 2021, https://nationalinterest.org/blog/buzz/compete-china-space-america-must -ramp-funding-186383.

13. Ibid.

14. Ibid.

15. Ibid.

16. "Joint Publication 3-14: Space Operations," U.S. Department of Defense, October 26, 2020,https://www.jcs.mil/Portals/36/Documents/Doctrine/pubs /jp3_14ch1.pdf; "Challenges to Security in Space," Defense Intelligence Agency,

January 2019,https://media.defense.gov/2019/Feb/11/2002088710/-1/-1/1/SPACE-SECURITY-CHALLENGES.PDF.

17. Bruce Cahan and Mir H. Sadat, "U.S. Space Policies for the New Space Age: Competing on the Final Economic Frontier," NewSpace New Mexico, January 6, 2021, https://www.politico.com/f/?id=00000177-9349-d713-a777-d7cfce4b0000; Stephen M. McCall, "Challenges to the United States in Space," Congressional Research Service, January 27, 2020, https://crsreports.congress.gov/product/pdf/IF/IF10337.

18. University of Illinois at Chicago Nathalie P. Voorhees Center for Neighborhood and Community Improvement & NASA, "National Aeronautics and Space Administration & Moon to Mars Program Economic Impact Study," August 2020, https://www.nasa.gov/sites/default/files/atoms/files/nasa_economic_impact_study.pdf.

19. Daniel Morgan, "NASA Appropriations and Authorizations: A Fact Sheet," Congressional Research Service, March 29, 2022, https://crsreports.congress.gov/product/pdf/R/R43419.

20. University of Illinois at Chicago Nathalie P. Voorhees Center for Neighborhood and Community Improvement & NASA, "National Aeronautics and Space Administration & Moon to Mars Program Economic Impact Study."

21. "NASA Report Details How Agency Significantly Benefits US Economy," NASA, September 25, 2020, https://www.nasa.gov/press-release/nasa-report-details-how-agency-significantly-benefits-us-economy.

22. "40 Years of NASA Spinoff," National Aeronautics and Space Administration, https://www.nasa.gov/offices/oct/40-years-of-nasa-spinoff.

23. "GPS Celebrates 25th Year of Operation," U.S. Space Force, April 27, 2020, https://www.spaceforce.mil/News/article/2166101/gps-celebrates-25th-year-of-operation/; Kathleen McTigue, "Economic Benefits of the Global Positioning System to the U.S. Private Sector Study," National Institute of Standards and Technology, October 2, 2019, https://www.nist.gov/news-events/news/2019/10/economic-benefits-global-positioning-system-us-private-sector-study.

24. "Protecting America's Global Positioning System," U.S. Department of Defense, https://www.defense.gov/explore/spotlight/protecting-gps/.

25. Greg Autry, "Space Research Can Save the Planet—Again," *Foreign Policy*, July 20, 2019, https://foreignpolicy.com/2019/07/20/space-research-can-save-the-planet-again-climate-change-environment/. Power grid synchronization is also reliant on GPS systems. See, for example, "Sync Up: Precision Timing for Smart Grids," Smart Energy International, September 27, 2021, https://www.smart-energy.com/industry-sectors/smart-grid/sync-up-precision-timing-for-smart-grids/.

26. McCall, "Challenges to the United States in Space."

27. Ibid.

28. Namrata Goswami, "The Economic and Military Impact of China's BeiDou Navigation System," *The Diplomat*, July 1, 2020, https://thediplomat.com/2020/07/the-economic-and-military-impact-of-chinas-beidou-navigation-system/.

29. On China's military aerospace sector activity, see, for example, Scott Harold, "Defeat, Not Merely Compete: China's Views of Its Military Aerospace Goals and Requirements in Relations to the US," RAND Corporation, 2018, https://www.rand.org/pubs/research_reports/RR2588.html.

30. Robert C. O'Brien, "The Chinese Communist Party's Ideology and Global Ambitions," National Security Adviser Speech in Phoenix, Arizona, June 24, 2020, https://trumpwhitehouse.archives.gov/briefings-statements/chinese-communist-partys-ideology-global-ambitions/; Christopher Wray, "The Threat Posed by the Chinese Government and the Chinese Communist Party to the Economic and National Security of the United States," Speech at Hudson Institute, Washington, DC, July 7, 2020, https://www.fbi.gov/news/speeches/the-threat-posed-by-the-chinese-government-and-the-chinese-communist-party-to-the-economic-and-national-security-of-the-united-states.

31. Central Committee of the Communist Party of China and the PRC State Council, "Outline of the National Innovation-Driven Development Strategy," Xinhua News Agency, May 19, 2016, http://www.xinhuanet.com/politics/2016-05/19/c_1118898033.htm (translated by Etcetera Language Group Inc. and edited by Ben Murphy, Center for Security and Emerging Technology, Georgetown University, https://cset.georgetown.edu/wp-content/uploads/t0076_innovation_driven_development_strategy_EN.pdf). The importance of space for the PRC is also outlined in the 2019 defense white paper "China's National Defense in the New Era," State Council Information Office of the People's Republic of China, July 2019, http://www.chinadaily.com.cn/specials/whitepaperonnationaldefensein newera.pdf.

32. Michael S. Chase, "The Space and Cyberspace Components of the Belt and Road Initiative," NBR Special Report no. 80, September 3, 2019, https://www.nbr.org/publication/the-space-and-cyberspace-components-of-the-belt-and-road-initiative/.

33. Mark Stokes, Gabriel Alvarado, Emily Weinstein, and Ian Easton, "China's Space and Counterspace Capabilities and Activities," Prepared for the U.S.-China Economic and Security Review Commission, March 30, 2020, https://www.uscc.gov/sites/default/files/2020-05/China_Space_and_Counterspace_Activities.pdf.

34. "China Sends 'World's First 6G' Test Satellite Into Orbit," BBC News, November 7, 2020, https://www.bbc.com/news/av/world-asia-china-54852131.

35. Namrata Goswami, Statement Before the U.S.-China Economic and Security Review Commission Hearing on "China in Space: A Strategic Competition?" April 25, 2019, 11, https://www.uscc.gov/sites/default/files/Namrata%20Goswami%20USCC%2025%20April.pdf.

36. Cao Siqi, "China Mulls $10 Trillion Earth-Moon Economic Zone," Global Times, November 1, 2019, https://www.globaltimes.cn/content/1168698.shtml.

37. Christopher Stone, "Hypersonic Weapons and the Case for a Space Tracking Layer," Defense Technology Program Brief No. 21, American Foreign Policy Council, December 2020, https://www.afpc.org/uploads/documents/Defense_Technology_Briefing_-_Issue_21.pdf; Christopher Stone, "Orbital Vigilance: The Need for Enhanced Space-Based Missile Warning and Tracking," Mitchell Institute, June 7, 2022, https://mitchellaerospacepower.org/wp-content/uploads/2022/06/Space_Based_Early_Warning_Policy_Paper_36-FINAL.pdf.

38. Lamont Colucci, "A Space Service in Support of American Grand Strategy," Space Review, February, 25, 2019, http://lamontcolucci.org/2019/02/25/the-space-review-a-space-service-in-support-of-american-grand-strategy/.

39. The White House, "National Space Policy of the United States of America," June 28, 2010, https://obamawhitehouse.archives.gov/sites/default/files/national_space_policy_6-28-10.pdf.

40. "National Security Space Strategy," Unclassified Summary, U.S. Department of Defense and Intelligence Community, January 2011, https://dod.defense.gov/Portals/1/features/defenseReviews/NSSS/NationalSecuritySpaceStrategyUnclassifiedSummary_Jan2011.pdf.

41. "Defense Space Strategy Summary," U.S. Department of Defense, June 2020, https://media.defense.gov/2020/Jun/17/2002317391/-1/-1/1/2020_DEFENSE_SPACE_STRATEGY_SUMMARY.PDF.

42. The White House, "Presidential Executive Order on Reviving the National Space Council," June 30, 2017, https://trumpwhitehouse.archives.gov/presidential-actions/presidential-executive-order-reviving-national-space-council/.

43. The White House, "President Donald J. Trump Is Unveiling an America First National Space Strategy," March 23, 2018, https://2017-2021.state.gov/president-donald-j-trump-is-unveiling-an-america-first-national-space-strategy/index.html; "Presidential Memorandum on Reinvigorating America's Human Space Exploration Program," December 11, 2017, https://trumpwhitehouse.archives.gov/presidential-actions/presidential-memorandum-reinvigorating-americas-human-space-exploration-program/; "Space Policy Directive-2, Streamlining Regulations on Commercial Use of Space," May 24, 2018, https://trumpwhitehouse.archives.gov/presidential-actions/space-policy-directive-2-streamlining-regulations-commercial-use-space/; "Space Policy Directive-3, National Space Traffic Management Policy," June 18, 2018, https://trumpwhitehouse.archives.gov/presidential-actions/space-policy-directive-3-national-space-traffic-management-policy/; "Text of Space Policy Directive-4: Establishment of the United States Space Force," February 19, 2019, https://media.defense.gov/2019/Mar/01/2002095015/-1/-1/1/SPACE-POLICY-DIRECTIVE-4-FINAL.PDF.

44. President Donald J. Trump, Address Before a Joint Session of the Congress on the State of the Union, February 4, 2020, https://www.govinfo.gov/content/pkg/DCPD-202000058/html/DCPD-202000058.htm; "Artemis Humanity's Return to the Moon," NASA, https://www.nasa.gov/specials/artemis/.

45. "National Space Policy of the United States of America."

46. "A New Era for Deep Space Exploration and Development," The White House National Space Council, July 23, 2020, https://aerospace.org/sites/default/files/2020-07/NSpC%20New%20Era%20for%20Space%2023Jul20.pdf.

47. Marcia Smith, "White House: Space Is One Area Where Biden and Trump Agree," SpacePolicyOnline.com, March 30, 2021, https://spacepolicyonline.com/news/white-house-space-is-one-area-where-biden-and-trump-agree/; Jen Psaki, "White House Daily Briefing," C-SPAN, March 30, 2021, https://www.c-span.org/video/?510369-1/white-house-press-secretary-covid-19-report-lacks-crucial-data; Marc Etkind and Jackie McGuinness, "Acting NASA Administrator Statement on Agency FY 2022 Discretionary Request," NASA, April 9, 2021, https://www.nasa.gov/press-release/acting-nasa-administrator-statement-on-agency-fy-2022-discretionary-request.

48. "Space Launch System Fact Sheet," NASA, https://www.nasa.gov/exploration/systems/sls/fs/sls.html.

49. Eric Berger, "Finally, We Know Production Costs for SLS and Orion, and They're Wild," *Ars Technica*, March 1, 2022, https://arstechnica.com /science/2022/03/nasa-inspector-general-says-sls-costs-are-unsustainable/; "Keeping Our Sights On Mars. Part 3: A Status Update and Review of NASA's Artemis Initiative," Testimony Before the House of Representatives Subcommittee on Space and Aeronautics, Committee on Science, Space, and Technology, National Aeronautics and Space Administration, Office of the Inspector General, January 20, 2022, https://science.house.gov/imo/media/doc/Martin%20Testimony4.pdf.

50. Micah Maidenberg, "SpaceX's Elon Musk Expects Starship to Deliver Launches at Lower Costs," *Wall Street Journal*, February 10, 2022, https://www .wsj.com/articles/elon-musk-expects-starship-to-deliver-launches-at-a-fraction -of-current-costs-11644549926.

51. Peter Garretson, interview with George Pullen, "George Pullen: A Four Trillion Dollar Space Economy," *Space Strategy*, podcast audio, November 24, 2021, https://anchor.fm/afpcspacepod/episodes/25--George-Pullen-A-Four-Trillion -Dollar-Space-Economy-e1aosj9.

52. Tina Highfill, Annabel Jouard, and Connor Franks, "Preliminary Estimates of the U.S. Space Economy, 2012–2018," *Survey of Current Business*, Bureau of Economic Analysis, December 2020, https://apps.bea.gov/scb/2020/12-december /pdf/1220-space-economy.pdf.

53. "U.S. Space Economy Through 2050: A Summary Economic Assessment," Foundation for the Future, https://www.f4f.space/space-act.

54. Michael Burlingame, "Abraham Lincoln: Campaigns and Elections," University of Virginia—Miller Center, https://millercenter.org/president/lincoln /campaigns-and-elections.

55. "Abraham Lincoln and Union Pacific," Union Pacific, https://www.up .com/heritage/history/lincoln/index.htm; Steven J. Butow, Thomas Cooley, Eric Felt, and Joel B. Mozer, "State of the Space Industrial Base 2020," Department of Defense, July 2020, https://cdn.afresearchlab.com/wp-content /uploads/2020/07/03155537/State-of-the-Space-Industrial-Base-2020-Report _July-2020_FINAL.pdf.

56. Ibid.

57. "Completion of the Transcontinental Railroad," Library of Congress, https://guides.loc.gov/this-month-in-business-history/may/completion -transcontinental-railroad.

58. "Landmark Legislation: The Pacific Railway Act of 1862," U.S. Senate, https://www.senate.gov/artandhistory/history/common/generic/PacificRailway Actof1862.htm.

59. "Completion of the Transcontinental Railroad."

60. Butow et al. "State of the Space Industrial Base 2020."

61. Ibid.

62. "Today in History—December 17, First Flight," Library of Congress, https:// www.loc.gov/item/today-in-history/december-17/.

63. Tim Sharp, "World's First Commercial Airline | The Greatest Moments in Flight," Space.com, May 22, 2018, https://www.space.com/16657-worlds-first -commercial-airline-the-greatest-moments-in-flight.html; "The Early Years of Air

Transportation, 1914–1927—The World's First Scheduled Airline," Smithsonian National Air and Space Museum, https://airandspace.si.edu/air-mail.

64. Sharp, "World's First Commercial Airline | The Greatest Moments in Flight."

65. "Dayton, Aviation, and the First World War," National Park Service, https://www.nps.gov/articles/dayton-aviation-and-the-first-world-war.htm.

66. Ibid.

67. "Airmail: A Brief History," U.S. Postal Service, https://about.usps.com/who-we-are/postal-history/airmail.pdf.

68. Ibid.

69. Ibid.

70. "Mobilization: The U.S. Army in World War II—The 50th Anniversary," U.S. Army, https://history.army.mil/documents/mobpam.htm; "War Production," PBS, https://www.pbs.org/kenburns/the-war/war-production/.

71. "Documents Related to FDR and Churchill—Background," National Archives, https://www.archives.gov/education/lessons/fdr-churchill.

72. "A History of WW2 in 25 Airplanes," *Smithsonian*, https://www.smithsonianmag.com/air-space-magazine/history-ww2-25-airplanes-180954056/; "Out-Producing the Enemy," National WWII Museum, https://www.nationalww2museum.org/sites/default/files/2017-07/mv-education-package.pdf; "World War 2 Fighter Aircraft," https://ww2f.org/war.html.

73. "Commercial Aviation at Mid-Century, 1941–1958," Smithsonian National Air and Space Museum, https://airandspace.si.edu/commercial-aviation-mid-century.

74. Ibid.

75. Erick Burgueño Salas, "Employment in the U.S. Aviation Industry—Statistics & Facts," Statista, April 19, 2022, https://www.statista.com/topics/4254/employment-in-the-us-aviation-industry/; "The Economic Impact of Civil Aviation on the U.S. Economy—State Supplement," U.S. Department of Transportation—Federal Aviation Administration, November 2020, https://www.faa.gov/sites/faa.gov/files/about/plans_reports/2020_nov_economic_impact_report.pdf.

76. On the complexities of coordinating the numerous government agencies necessary for a sustained moon presence, see Thomas D. Olszewski and Jericho Locke, "Potential Roles of Federal Agencies in Creating a Sustainable Presence on the Moon," IDA Science and Technology Policy Institute, March 2020, https://www.ida.org/-/media/feature/publications/p/po/potential-roles-of-federal-agencies-in-creating-a-sustainable-presence-on-the-moon/p-12071.ashx%C2%A0.

77. H.R.6395—William M. (Mac) Thornberry National Defense Authorization Act for Fiscal Year 2021, https://www.congress.gov/bill/116th-congress/house-bill/6395/text/enr.

78. Ibid.

79. Peter Garretson, interview with Rick Tomlinson, "A Far Reaching Vision for Space," *Space Strategy*, podcast audio, April 14, 2021, https://anchor.fm/afpcspacepod/episodes/A-Far-Reaching-Vision-for-Space-euu7d2.

80. Peter Garretson, interview with Brent Ziarnick, "'A Spacepower Marathon: Maximizing Total National Spacepower': A Deep Dive Into Asteroid Mining & Space Logistics," *Space Strategy*, podcast audio, April 28, 2021, https://anchor.fm/afpcspacepod/episodes/A-Spacepower-Marathon-Maximizing-Total-National-Spacepower-evqfr1.

81. Peter Garretson, interview with Joel Sercel, "'There Is Gold in Them Hills!': A Deep Dive Into Asteroid Mining & Space Logistics," *Space Strategy*, podcast audio, June 16, 2021, https://anchor.fm/afpcspacepod/episodes/There-is-Gold-in-Them-Hills--A-Deep-Dive-Into-Asteroid-Mining--Space-Logistics-e12tcfd.

82. Peter Garretson, interview with Steven Butow, "'Think The Expanse': How DIU Is Enabling the Nation to Do 'Interesting Things' in Space," *Space Strategy*, podcast audio, June 1, 2021, https://anchor.fm/afpcspacepod/episodes/Think-The-Expanse-How-DIU-is-enabling-the-nation-to-do-interesting-things-in-Space-e1205hh.

83. Peter Garretson, interview with General James Cartwright, "'A Domain for Commerce': Moving From a Discovery Architecture to a Sustained and Commerce-Centric Architecture," *Space Strategy*, podcast audio, June 24, 2021, https://anchor.fm/afpcspacepod/episodes/A-Domain-for-Commerce-Moving-from-a-discovery-architecture-to-a-sustained-and-commerce-centric-architecture-e13cjt2.

84. Peter Garretson, interview with Robert Zubrin, "Our World Comes With an Infinite Sky—Mars Beckons," *Space Strategy*, podcast audio, April 21, 2021, https://anchor.fm/afpcspacepod/episodes/5--Bob-Zubrin-Our-World-Comes-with-an-Infinite-Sky----Mars-Beckons-evd54k.

CHAPTER 5

1. Mackenzie Eaglen, "Abort the Space Corps on the Launchpad," *Breaking Defense*, July 13, 2017, https://breakingdefense.com/2017/07/abort-the-space-corps-on-the-launchpad/.

2. William D. Hartung and Mandy Smithberger, "Does America Need a Space Force?" *Defense One*, October 9, 2019, https://www.defenseone.com/ideas/2019/10/does-america-need-space-force/160475/.

3. Todd Harrison, "Why We Need a Space Force," Center for Strategic and International Studies, October 3, 2018, https://www.csis.org/analysis/why-we-need-space-force.

4. "United States Space Force, Size and Cost," https://www.spaceforce.mil/About-Us/FAQs/Whats-the-Space-Force/.

5. U.S. Department of Defense, "Operations and Maintenance Programs (O-1) Revolving and Management Funds (RF-1)," Department of Defense Budget FY 2023, April 2022, https://comptroller.defense.gov/Portals/45/Documents/defbudget/FY2023/FY2023_o1.pdf.

6. Congressman Mike Rogers, Chairman, House Armed Services Strategic Forces Subcommittee, "Remarks to the 2017 Space Symposium," *Strategic Studies Quarterly* 11, no. 2 (Summer 2017), https://www.airuniversity.af.edu/Portals/10/SSQ/documents/Volume-11_Issue-2/Rogers.pdf.

7. Rebecca Kheel, "Space Corps Proponents: 'The Time for Study Is Over,'" *The Hill*, July 19, 2017, https://thehill.com/policy/defense/342811-space-corps-proponents-time-for-study-is-over/; "Defense Space Acquisitions: Too Early to Determine If Recent Changes Will Resolve Persistent Fragmentation in Management and Oversight," U.S. Government Accountability Office, July 27, 2016, https://www.gao.gov/assets/gao-16-592r.pdf.

8. Sandra Erwin, "Congressman Rogers: A Space Corps Is 'Inevitable,'" *SpaceNews*, December 2, 2017, https://spacenews.com/congressman-rogers-a-space-corps-is-inevitable/.

9. David Thornton, "Intelligence Agencies Need to Be More Aggressive, Says Former Lawmaker," *Federal News Network*, March 31, 2016, https://federalnewsnetwork.com/cybersecurity/2016/03/intelligence-agencies-need-aggressive-says-former-lawmaker/.

10. Tim Ventura, interview with Peter Garretson, "The Battle for the Soul of Space Force," Medium.com, May 1, 2020, https://medium.com/predict/the-battle-for-the-soul-of-the-space-force-e9e9c3d84c8a.

11. John Hattendorf, *Mahan On Naval Strategy* (Annapolis, MD: Naval Institute Press, 1991), 24.

12. U.S. Department of Defense, "Competition Continuum," Joint Doctrine Note 1-19, June 3, 2019, https://www.jcs.mil/Portals/36/Documents/Doctrine/jdn_jg/jdn1_19.pdf.

13. The White House, "Space Policy Directive-4," February 19, 2019, https://media.defense.gov/2019/Mar/01/2002095015/-1/-1/1/SPACE-POLICY-DIRECTIVE-4-FINAL.PDF.

14. "S.1790—National Defense Authorization Act for Fiscal Year 2020," https://www.congress.gov/bill/116th-congress/senate-bill/1790.

15. U.S. Space Force, "Spacepower: Doctrine for Space Forces," Space Capstone Publication, June 2020, https://www.spaceforce.mil/Portals/1/Space Capstone Publication_10 Aug 2020.pdf.

16. U.S. Space Force, "Mission," https://www.spaceforce.mil/About-Us/FAQs/Whats-the-Space-Force/.

17. U.S. Department of Defense, "Space Operations," Joint Publication 3-14, April 10, 2018, Incorporating Change 1, October, 26, 2020, https://www.jcs.mil/Portals/36/Documents/Doctrine/pubs/jp3_14ch1.pdf.

18. John E. Shaw, Jean Purgason, and Amy Soileau, "Sailing the New Wine-Dark Sea: Space as a Military Area of Responsibility," *Æther: A Journal of Strategic Airpower and Spacepower* 1, no. 1 (Spring 2022); Ramin Skibba, "Space Command's Lt. Gen John Shaw Says Space Is 'Under Threat,'" Wired.com, April 15, 2022, https://www.wired.com/story/space-commands-lt-gen-john-shaw-on-the-future-of-space-security/; James Dickenson, U.S. Space Command, "Fiscal Year 2023 Priorities and Posture of United States Space Command," presentation to the U.S. Senate Armed Services Committee, March 1, 2022, https://www.armed-services.senate.gov/imo/media/doc/USSPACECOM%20FY23%20Posture%20Statement%20SASC%20FINAL.pdf.

19. Carl A. Poole and Robert A. Bettinger, "Black Space Versus Blue Space: A Proposed Dichotomy of Future Space Operations," *Air & Space Power Journal* 35, no. 1 (Spring 2021), https://www.airuniversity.af.edu/Portals/10/ASPJ/journals/Volume-35_Issue-1/F-Poole.pdf.

20. U.S. Department of Defense, "U.S. Space Force," February 2019, https://media.defense.gov/2019/Mar/01/2002095012/-1/-1/1/UNITED-STATES-SPACE-FORCE-STRATEGIC-OVERVIEW.PDF

21. Ibid.

22. Brent Ziarnick, "A Practical Guide for Spacepower Strategy," *Space Force Journal*, Issue 1, January 31, 2021, https://spaceforcejournal.org/a-practical

-guide-for-spacepower-strategy/. According to an article by LTC Brad Townsend, "Space Power and the Foundations of an Independent Space Force," https:// www.airuniversity.af.edu/Portals/10/ASPJ/journals/Volume-33_Issue-4/F -Townsend.pdf, the 2009 version of JP 3-14 defines spacepower as "the total strength of a nation's capabilities to conduct and influence activities to, in, through, and from space to achieve its objectives."

23. Ibid.

24. Peter Garretson, interview with Rick Tumlinson, "A Far-Reaching Vision for Space," *Space Strategy*, podcast audio, April 14, 2021, https://anchor.fm /afpcspacepod/episodes/A-Far-Reaching-Vision-for-Space-euu7d2.

25. Ziarnick, "A Practical Guide for Spacepower Strategy."

26. Patrick M. Cronin, Masashi Murano, and H. R. McMaster, "Transcript: U.S. Space Strategy and Indo-Pacific Cooperation," Hudson Institute, November 15, 2019, https://www.hudson.org/research/15481-transcript-u-s-space-strategy -and-indo-pacific-cooperation.

27. Tim Venura, "The Space Force Is a Transformative Victory for American Leadership in the 21st Century," *Predict*, July 26, 2020, https://medium.com/predict /the-space-force-is-a-transformative-victory-for-american-leadership-in-the-21st -century-61f2e16a7fa5.

28. Peter Garretson, "A Historic National Vision for Spacepower," *War on the Rocks*, September 9, 2019, https://warontherocks.com/2019/09/a-historic -national-vision-for-spacepower/.

29. Valerie Insinna, "The Pentagon Has Sent a New Legislative Proposal on the Space Force to Congress," *Defense News*, March 19, 2020, https://www.defensenews .com/space/2020/03/19/the-pentagon-has-sent-a-new-legislative-proposal-on -the-space-force-to-congress/.

30. Spencer Kaplan, "Eyes on the Prize: The Strategic Implications of Cislunar Space and the Moon," Aerospace Security Project, Center for Strategic and International Studies, July 13, 2020, http://aerospace.csis.org/wp-content /uploads/2020/07/20200714_Kaplan_Cislunar_FINAL.pdf.

31. Theresa Hitchens, "Exclusive: Pentagon Poised to Unveil, Demonstrate Classified Space Weapon," *Breaking Defense*, August 20, 2021, https://breakingdefense .com/2021/08/pentagon-posed-to-unveil-classified-space-weapon/.

32. Brian Weeden and Victoria Samson, "Global Counterspace Capabilities," Secure World Foundation, April 2021, https://swfound.org/media/207162/swf _global_counterspace_capabilities_2021.pdf.

33. Ibid.

34. Joseph Trevithick, "Space Force Just Received Its First New Offensive Weapon," *The Drive*, March 13, 2020, https://www.thedrive.com/the-war-zone/32570 /space-force-just-received-its-first-new-offensive-weapon.

35. U.S. Space Force, "Spacepower: Doctrine for Space Forces."

36. Ibid.

37. Ibid.

38. Ibid.

39. For example, on Vice President Harris's announcement that the United States commits not to conduct destructive, direct-ascent anti-satellite (ASAT) missile testing (https://www.whitehouse.gov/briefing-room/statements-releases /2022/04/18/fact-sheet-vice-president-harris-advances-national-security-norms

-in-space/), Gerrit Dalman commented on LinkedIn on April 20, 2022, "This is a legitimate policy. However, I am concerned that our response to the irresponsible use of violence is to criticize not counter. In no other domain would we allow a competitor to brazenly demonstrate capabilities and deter through shame without the backing an in-kind competing capability" (https://www.linkedin.com /feed/update/urn:li:activity:6922223971765542912/). Joshua Carlson, author of *Spacepower Ascendant: Space Development Theory and a New Space Strategy* (https:// www.amazon.com/Spacepower-Ascendant-Development-Theory-Strategy/dp /B08BWGPR8V), followed up on Dalman's comments, adding, "I've never believed that the right way to deal with revisionist powers is to disarm yourself. They want to revise the world system, and you make it easier by de-weaponizing yourself" (https://www.linkedin.com/feed/update/urn:li:activity:6922223971765542912/).

40. Hitchens, "Exclusive: Pentagon Poised to Unveil, Demonstrate Classified Space Weapon."

41. Ibid.

42. For example, regarding the ability to deter China, read Krista Langeland and Derek Grossman, "Tailoring Deterrence for China in Space," RAND Corporation, March 2021, https://www.rand.org/pubs/research _reports/RRA943-1.html; see also John Klein, "Towards a Better U.S. Space Strategy: Addressing the Strategy Mismatch," Strategy Bridge, September 9, 2019, https://thestrategybridge.org/the-bridge/2019/9/9/towards-a-better-us-space -strategy-addressing-the-strategy-mismatch.

43. Hitchens, "Exclusive: Pentagon Poised to Unveil, Demonstrate Classified Space Weapon."

44. Erin Salinas, "Space Situational Awareness Is Space Battle Management," Air Force Space Command, May 16, 2018, https://www.afspc.af.mil /News/Article-Display/Article/1523196/space-situational-awareness-is-space -battle-management/.

45. Sandra Erwin, "Air Force: SSA Is No More; It's 'Space Domain Awareness,'" *SpaceNews*, November 14, 2019, https://spacenews.com/air-force-ssa-is-no-more -its-space-domain-awareness/.

46. Ibid.

47. Peter Garretson, interview with Steven Butow, "'Think The Expanse': How DIU Is Enabling the Nation to Do 'Interesting Things' in Space," *Space Strategy*, podcast audio, June 1, 2021, https://anchor.fm/afpcspacepod/episodes/Think -The-Expanse-How-DIU-is-enabling-the-nation-to-do-interesting-things-in -Space-e1205hh.

48. Kaplan, "Eyes on the Prize"

49. Peter Garretson, interview with Bhavya Lal, "Incorporating the Solar System Into Our Economic Sphere"—The Relevance of Artemis and NASA to the National Strategy and National Security, *Space Strategy*, podcast audio, August 13, 2021, https://anchor.fm/afpcspacepod/episodes/Bhavya-Lal-Incorporating-the-Solar -System-into-Our-Economic-Sphere---The-Relevance-of-Artemis-and-NASA-to -National-Strategy-and-National-Security-e15rsdn.

50. Peter Garretson, interview with Victoria Samson, "Constructing International Norms for Space," *Space Strategy*, podcast audio, July 29, 2021, https://anchor.fm

/afpcspacepod/episodes/Victoria-Samson-Constructing-International-Norms
-for-Space-e157b2o.

51. Peter Garretson, interview with Dr. M. V. "Coyote" Smith, "Space Force
Mission Is a Better Tomorrow," *Space Strategy*, podcast audio, July 3, 2021,
https://anchor.fm/afpcspacepod/episodes/Dr--Coyote-Smith-Space-Force
-Mission-is-a-better-tomorrow-e13rpu3.

52. Garretson, interview with Rick Tumlinson.

53. Peter Garretson, "Space Force's Jupiter-Sized Culture Problem," *War on
the Rocks*, July 11, 2019, https://warontherocks.com/2019/07/space-forces
-jupiter-sized-culture-problem/.

54. Ibid.

55. Ibid.

56. Sandra Erwin, "Space Force Warned to Avoid Past Mistakes as It Pursues New
Satellite Acquisitions," *SpaceNews*, May 24, 2021, https://spacenews.com/space
-force-warned-to-avoid-past-mistakes-as-it-pursues-new-satellite-acquisitions/.

57. Ibid.

58. Ibid.

59. Rachel S. Cohen, "Congress Turns Down a Space National Guard Again, but
Space Force Isn't Giving Up," *Air Force Times*, December 29, 2021, https://www
.airforcetimes.com/news/your-air-force/2021/12/29/congress-turns-down-a
-space-national-guard-again-but-space-force-isnt-giving-up/.

60. The White House, "Statement of Administration Policy H.R. 4350—
National Defense Authorization Act for Fiscal Year 2022," Office of Management
and Budget, September 21, 2021, https://www.whitehouse.gov/wp-content
/uploads/2021/09/SAP-HR-4350.pdf.

61. Peter Garretson, "Establish Space National Guard Now," *The Hill*, June 7,
2022, https://thehill.com/opinion/national-security/3514195-establish-the-space
-national-guard-now/.

62. Steve Beynon, "Space Guard 'Among My Most Pressing Concerns,' National
Guard Chief Tells Congress," Military.com, May 4, 2021, https://www.military
.com/daily-news/2021/05/04/national-guard-chief-highlight-guards-space
-mission-congress.html.

63. Christopher Stone, "The Space National Guard Already Exists—Congress
Should Recognize It," *The Hill*, February 20, 2022, https://thehill.com/opinion
/national-security/594482-the-space-national-guard-already-exists-congress
-should-recognize.

64. Ibid.

65. Ibid.

66. Brent Ziarnick, "Space Force Reserve Too Important to Be Dictated by Active
Duty," *Air Force Times*, April 20, 2021, https://www.airforcetimes.com/opinion
/commentary/2021/04/20/space-force-reserve-too-important-to-be-dictated-by
-active-duty/.

67. John P. Roth, Charles Brown, and John Raymond, "Department of the Air
Force Posture Statement Fiscal Year 2022," https://docs.house.gov/meetings
/AP/AP02/20210507/112533/HHRG-117-AP02-Bio-RaymondJ-20210507.pdf.

68. Ibid.

69. Carl Poole and Robert A. Bettinger, "The Cosmic Sandbox: An Advocated Military Role in Future Space Commerce and Exploration," *Space Force Journal*, Issue 1, January 31, 2021, https://spaceforcejournal.org/the-cosmic-sandbox-an-advocated-military-role-in-future-space-commerce-and-exploration/.

70. U.S. Space Force, "Spacepower: Doctrine for Space Forces."

71. Bill Woolf, "What Can the Space Force Do for America?" American Foreign Policy Council Capitol Hill Briefing, April 12, 2021, https://www.afpc.org/news/events/afpc-capitol-hill-briefing-what-can-the-space-force-do-for-america.

72. Colin Clark, "NRO, NGA, SPACECOM, Space Force Hammer Out Boundaries," *Breaking Defense*, August 24, 2021, https://breakingdefense.com/2021/08/nro-nga-spacecom-space-force-hammer-out-boundaries/.

73. Ibid.

74. Roth et al., "Department of the Air Force Posture Statement Fiscal Year 2022."

75. Ibid.

76. Ibid.

77. Racheal Blodgett, "Our Missions and Values," NASA, August 5, 2021, https://www.nasa.gov/careers/our-mission-and-values.

78. "Memorandum of Understanding Between the National Aeronautics and Space Administration and the United States Space Force," NASA, September 20, 2021, https://www.nasa.gov/sites/default/files/atoms/files/nasa_ussf_mou_21_sep_20.pdf.

79. Roth et al., "Department of the Air Force Posture Statement Fiscal Year 2022."

80. Garretson, interview with Steven Butow; see also Peter Garretson, interview with Joel Sercel, "'There Is Gold in Them Hills!': A Deep Dive Into Asteroid Mining & Space Logistics," *Space Strategy*, podcast audio, June 16, 2021, https://anchor.fm/afpcspacepod/episodes/There-is-Gold-in-Them-Hills--A-Deep-Dive-Into-Asteroid-Mining--Space-Logistics-e12tcfd.

81. Garretson, interview with Rick Tumlinson; see also Peter Garretson, interview with Josh Carlson, "Explore-Expand-Exploit-Exclude: Space Development Theory and the Value of Vision," *Space Strategy*, podcast audio, May 21, 2021, https://anchor.fm/afpcspacepod/episodes/Explore-Expand-Exploit-Exclude-Space-Development-Theory-and-the-Value-of-Vision-e11bda3.

82. Peter Garretson, "The Purpose of a Space Force Is a Spacefaring Economy," *The Hill*, June 26, 2019, https://thehill.com/opinion/technology/450519-the-purpose-of-a-space-force-is-a-spacefaring-economy.

83. Peter Garretson, "The First Duty of a Space Force Is to Protect Space Commerce," *Politico*, June 21, 2019, https://www.politico.com/story/2019/06/21/opinion-space-force-commerce-1374229.

84. For information on the competition of resources in space, see Namrata Goswami and Peter A. Garretson, *Scramble for the Skies: The Great Power Competition to Control the Resources of Outer Space* (Lanham, MD: Lexington Books, 2020).

85. Poole and Bettinger, "The Cosmic Sandbox."

86. Ibid.

87. Garretson, interview with Josh Carlson.

88. Poole and Bettinger, "The Cosmic Sandbox"; see also Garretson, interview with Dr. M. V. "Coyote" Smith.

89. Garretson, interview with Dr. M. V. "Coyote" Smith.

90. Peter Garretson, interview with Kara Cunzeman, "Abundance, Aspiration, and Strategic Foresight," *Space Strategy*, podcast audio, May 12, 2021, https://anchor.fm/afpcspacepod/episodes/Abundance--Aspiration--and-Strategic-Foresight-e10ngnu.

91. Fred Kennedy, "It's Time to Equip the U.S. Space Force With the Ability to Project Force," *Forbes*, July 22, 2020, https://www.forbes.com/sites/fredkennedy/2020/07/22/its-time-to-equip-the-us-space-force-with-the-ability-to-project-force/.

92. Garretson, interview with Rick Tumlinson.

93. Jeff Becker, "A Starcruiser for Space Force: Thinking Through the Imminent Transformation of Spacepower," *War on the Rocks*, May 19, 2021, https://warontherocks.com/2021/05/a-starcruiser-for-space-force-thinking-through-the-imminent-transformation-of-spacepower/.

94. Mark Whittington, "Is the Space Force About to Acquire SpaceX Starships?" *The Hill*, June 6, 2021, https://thehill.com/opinion/technology/557026-is-the-space-force-about-to-acquire-spacex-starships.

95. Becker, "A Starcruiser for Space Force."

96. Ibid.

97. Garretson, interview with Josh Carlson.

98. Woolf, "What Can the Space Force Do for America?"

99. Garretson, interview with Dr. M. V. "Coyote" Smith.

100. Peter Garretson, interview with James Cartwright, "'A Domain for Commerce': Moving From a Discovery Architecture to a Sustained and Commerce-Centric Architecture," *Space Strategy*, podcast audio, June 24, 2021, https://anchor.fm/afpcspacepod/episodes/A-Domain-for-Commerce-Moving-from-a-discovery-architecture-to-a-sustained-and-commerce-centric-architecture-e13cjt2.

101. Peter Garretson, "What Can the Space Force Do for America?" American Foreign Policy Council Capitol Hill Briefing, April 12, 2021, https://www.afpc.org/news/events/afpc-capitol-hill-briefing-what-can-the-space-force-do-for-america.

CHAPTER 6

1. "2019 Report to Congress," U.S.-China Economic and Security Review Commission, November 2019, https://www.uscc.gov/sites/default/files/2019-11/2019%20Annual%20Report%20to%20Congress.pdf.

2. Spencer Kaplan, "Eyes on the Prize: The Strategic Implications of Cislunar Space and the Moon," Aerospace Security Project, Center for Strategic and International Studies, July 13, 2020, http://aerospace.csis.org/wp-content/uploads/2020/07/20200714_Kaplan_Cislunar_FINAL.pdf.

3. Stephen Chen, "China Speeds Up Moon Base Plan in Space Race Against the US," *South China Morning Post*, December 29, 2021, https://www.scmp.com/news/china/science/article/3161324/china-speeds-moon-base-plan-space-race-against-us.

4. Clementine G. Starling, Mark J. Massa, Christopher P. Mulder, and Julia T. Siegel, "The Future of Security in Space: A Thirty-Year US Strategy," Atlantic Council Strategy Paper Series, April 11, 2021, https://www.atlanticcouncil.org/content-series/atlantic-council-strategy-paper-series/the-future-of-security-in-space/.

5. Loren Grush, "China Complains to UN After Maneuvering Its Space Station Away From SpaceX Starlink Satellites," *The Verge*, December 28, 2021, https://www.theverge.com/2021/12/28/22857035/china-spacex-starlink-tianhe-space-station-satellites-collisions.

6. Antony Blinken, "Russia Conducts Destructive Anti-Satellite Missile Test," Press Statement, U.S. Department of State, November 15, 2021, https://www.state.gov/russia-conducts-destructive-anti-satellite-missile-test/.

7. The White House, "Fact Sheet: Vice President Harris Advances National Security Norms in Space," April 18, 2022, https://www.whitehouse.gov/briefing-room/statements-releases/2022/04/18/fact-sheet-vice-president-harris-advances-national-security-norms-in-space/.

8. Lloyd Austin, "Tenets of Responsible Behavior in Space," Memorandum for Secretaries of the Military Departments, Office of the Secretary of Defense, July 7, 2021, https://media.defense.gov/2021/Jul/23/2002809598/-1/-1/0/TENETS-OF-RESPONSIBLE-BEHAVIOR-IN-SPACE.PDF.

9. Ibid.

10. Theresa Hitchens, "Exclusive: In a First, SecDef Pledges DoD to Space Norms," *Breaking Defense*, July 19, 2021, https://breakingdefense.com/2021/07/exclusive-in-a-first-secdef-pledges-dod-to-space-norms/.

11. Mike Wall, "Kessler Syndrome and the Space Debris Problem," Space.com, November 15, 2021, https://www.space.com/kessler-syndrome-space-debris.

12. Wayne White, "Salvage Law for Outer Space," Engineering, Construction, and Operations In Space III, SPACE '92, Proceedings of the Third International Conference, Denver, CO, May 31–June 4, 1992, https://www.academia.edu/6983159/Salvage_Law_for_Outer_Space.

13. Peter Garretson, interview with Joel Sercel, "'There Is Gold in Them Hills!': A Deep Dive Into Asteroid Mining & Space Logistics," *Space Strategy*, podcast audio, June 16, 2021, https://anchor.fm/afpcspacepod/episodes/There-is-Gold-in-Them-Hills--A-Deep-Dive-Into-Asteroid-Mining--Space-Logistics-e12tcfd.

14. Peter Garretson, Alfred B. Anzaldúa, and Hoyt Davidson, "Catalyzing Space Debris Removal, Salvage, and Use," *Space Review*, December 9, 2019, https://www.thespacereview.com/article/3847/1.

15. Adam Mann and Tereza Pultarova, "Starlink: SpaceX's Satellite Internet Project," Space.com, January 7, 2022, https://www.space.com/spacex-starlink-satellites.html.

16. Marques Brownlee, "The New Space Race!" WaveForm: The MKBHD Podcast, September 24, 2021, https://podcasts.apple.com/us/podcast/the-new-space-race/id1474429475?i=1000536479977.

17. Peter Garretson, interview with Theresa Hitchens, "Space Resources, Space Militarization, Weapons, Norms, SciFi, Humans as a Migratory Species & the Expanse as a Cautionary Tale," *Space Strategy*, podcast audio, October 30, 2021, https://anchor.fm/afpcspacepod/episodes/22--Theresa-Hitchens-Space-Resources--Space-Militarization--Weapons--Norms--SciFi--Humans-as-a-Migratory-Species--The-Expanse-as-a-Cautionary-Tale-e19jhqi.

18. Peter Garretson, interview with Wayne White, "The Space Pioneer Act," *Space Strategy*, podcast audio, January 17, 2022, https://anchor.fm/afpcspacepod/episodes/30--Wayne-White-The-Space-Pioneer-Act-e1d232a.

19. Peter Garretson, interview with Victoria Samson, "Constructing International Norms for Space," *Space Strategy*, podcast audio, July 29, 2021, https://anchor.fm/afpcspacepod/episodes/16--Victoria-Samson-Constructing-International-Norms-for-Space-e157b2o.

20. Ibid.

21. Hoss Cartwright and Deborah Lee James, "The Space Rush: New US Strategy Must Bring Order, Regulation," *Breaking Defense*, March 26, 2021, https://breakingdefense.com/2021/03/the-space-rush-new-us-strategy-must-bring-order-regulation/.

22. Christopher R. Moore et al., "Sediment Cores from White Pond, South Carolina, Contain a Platinum Anomaly, Pyrogenic Carbon Peak, and Coprophilous Spore Decline at 12.8 ka," *Nature*, October 22, 2019, https://www.nature.com/articles/s41598-019-51552-8.

23. "Guangzhou Science and Technology Forum: China's Space Science and Technology to Build a Dream of the Sky (广州科普大讲坛：中国航天科技筑梦天疆)," Guangzhou Science and Technology Association, July 30, 2021, https://www.163.com/dy/article/GG5KVJ540514QQM0.html.

24. "Planetary Defense! NASA Intends to Allow Spacecraft to Intercept Asteroids and Save the Earth by Changing Its Orbit (行星防御！NASA拟让飞船拦截小行星通过改变其轨道拯救地球)," *Xinhua*, October 16, 2020, http://www.xinhuanet.com/tech/2020-10/16/c_1126617533.htm.

25. The State Council Information Office of the People's Republic of China, "Full Text: China's Space Program: A 2021 Perspective," China National Space Administration, January 28, 2022, http://english.www.gov.cn/archive/whitepaper/202201/28/content_WS61f35b3dc6d09c94e48a467a.html.

26. "Double Asteroid Redirection Test (DART) Mission," NASA Planetary Defense, https://www.nasa.gov/planetarydefense/dart.

27. "Memorandum of Understanding Between the National Aeronautics and Space Administration and the United States Space Force," NASA, September 21, 2020, https://www.nasa.gov/sites/default/files/atoms/files/nasa_ussf_mou_21_sep_20.pdf; Ramin Skibba, "Space Command's Lt. Gen John Shaw Says Space Is 'Under Threat,'" Wired.com, April 15, 2022, https://www.wired.com/story/space-commands-lt-gen-john-shaw-on-the-future-of-space-security/.

28. Garretson, interview with Wayne White.

29. "2222 (XXI). Treaty on Principles Governing the Activities of States in the Exploration and Use of Outer Space, Including the Moon and Other Celestial Bodies," U.N. Office for Outer Space Affairs, 1499th plenary meeting, December 19, 1966, https://www.unoosa.org/oosa/en/ourwork/spacelaw/treaties/outerspacetreaty.html.

30. Peter Garretson, "What's in a Code? Putting Space Development First," *Space Review*, April 7, 2014, https://www.thespacereview.com/article/2487/1.

31. "The Artemis Accords: Principles for a Safe, Peaceful, and Prosperous Future," NASA, https://www.nasa.gov/specials/artemis-accords/index.html.

32. Country and global GDP estimates were generated using the International Monetary Fund database, available at https://www.imf.org/external/datamapper/NGDPD@WEO/OEMDC/ADVEC/WEOWORLD.

33. "France Becomes Twentieth Nation to Sign the Artemis Accords," U.S. Department of State, June 7, 2022, https://www.state.gov/france-becomes-twentieth -nation-to-sign-the-artemis-accords/.

34. Charlotte Mathieu, "Assessing Russia's Space Cooperation With China and India: Opportunities and Challenges for Europe," European Space Policy Institute, June 12, 2008, https://www.espi.or.at/reports/assessing-russias-space -cooperation-with-china-and-india/; Guo Xiaoqiang (郭晓琼), *China Russia Economic Cooperation in the New Era: New Trends and New Problems* (新时代俄中经贸合作： 新 i趋势与新问题) (Beijing: Chinese Academy of Social Sciences Press, 2021), 131.

35. Mathieu, "Assessing Russia's Space Cooperation with China and India: Opportunities and Challenges for Europe," 21; Mike Gruntman, "Joining the Club," in *Blazing the Trail: Early History of Spacecraft and Rocketry* (Reston, VA: American Institute of Aeronautics and Astronautics, 2004), excerpted by Gruntman at http://www.astronauticsnow.com/bttp/btt_p_440_441.pdf.

36. Mathieu, "Assessing Russia's Space Cooperation with China and India: Opportunities and Challenges for Europe," 82.

37. Almudena Azcárate Ortega, "Placement of Weapons in Outer Space: The Dichotomy Between Word and Deed," *Lawfare*, January 28, 2021, https://www .lawfareblog.com/placement-weapons-outer-space-dichotomy-between-word -and-deed.

38. Ibid.

39. Guo, *China Russia Economic Cooperation in the New Era*, 131.

40. "International Lunar Research Station Guide for Partnership (V1.0)," China National Space Administration, June 2021, http://www.cnsa.gov.cn/english /n6465652/n6465653/c6812150/content.html.

41. Namrata Goswami, "The Strategic Implications of the China-Russia Lunar Base Cooperation Agreement," *The Diplomat*, March 19, 2021, https://thediplomat .com/2021/03/the-strategic-implications-of-the-china-russia-lunar-base -cooperation-agreement/.

42. The quote from Jana Robinson was provided via email to Peter Garretson and Richard Harrison on July 8, 2022. For more information on her previous analysis, see Jana Robinson, Tereza B. Kupková, and Patrik Martínek, "Strategic Competition for Space Partnerships and Markets," in *Handbook of Space Security* (Cham: Springer, 2020), https://doi.org/10.1007/978-3-030-22786-9_141-2.

43. This assessment is based on satellite data from the Union of Concerned Scientist Database (updated January 1, 2022), https://www.ucsusa.org/resources /satellite-database.

44. Irene Klotz, "Burgeoning Satellite Industry Paving Way to $1 Trillion Space Economy," *Aviation Week*, August 24, 2021, https://aviationweek.com/aerospace /program-management/burgeoning-satellite-industry-paving-way-1-trillion -space-economy; "Beyond the Planet: Charting the Future of the Space Sector," International Telecommunication Union, December 14, 2020, https://www.itu .int/hub/2020/12/beyond-the-planet-charting-the-future-of-the-space-sector/.

45. Garretson, interview with Victoria Samson.

46. Aaron Mehta, "Increasing Allied Role in Space a 'Priority' for Space Command Head," *Defense News*, September 3, 2019, https://www.defensenews .com/space/2019/09/03/increasing-allied-role-in-space-a-priority-for-space -command-head/.

47. See, for example, the most recent joint declaration with Japan: Sean Potter, "NASA Administrator Signs Declaration of Intent With Japan on Artemis, Space Station Cooperation," NASA, July 13, 2020, https://www.nasa.gov/feature/nasa -administrator-signs-declaration-of-intent-with-japan-on-artemis-space-station.

48. Peter Garretson, interview with Henry Sokolski, "A China-U.S. War in Space: The After-Action Report," *Space Strategy*, podcast audio, February 3, 2022, https:// anchor.fm/afpcspacepod/episodes/32--Henry-Sokolski-A-China-U-S--war-in -space-The-after-action-report-e1dss4r.

49. See, for example, France's space strategy: "Florence Parly Unveils the French Space Defence Strategy," Permanent Representation of France to the Conference on Disarmament, July 25, 2019, https://cd-geneve.delegfrance.org /Florence-Parly-unveils-the-French-space-defence-strategy.

50. Michael Sheetz and Yelena Dzhanova, "Top Russian Space Official Dismisses NASA's Moon Plans, Considering a Lunar Base With China Instead," CNBC, July 15, 2020, https://www-cnbc-com.cdn.ampproject.org/c/s/www.cnbc.com /amp/2020/07/15/russia-space-chief-dmitry-rogozin-dismisses-nasas-Moon -program-considering-china-lunar-base.html.

51. "Australia Announces New 'Space Command' Defence Agency," *BBC News*, March 22, 2022, https://www.bbc.com/news/world-australia-60835136; Theresa Hitchens, "US, Close Allies Sign 'Call to Action' in Space Defense," *Breaking Defense*, February 22, 2022, https://breakingdefense.com/2022/02/us-close-allies-sign -call-to-action-in-space-defense/; Ajey Lele, "India Needs Its Own Space Force," *SpaceNews*, May 28, 2019, https://spacenews.com/op-ed-india-needs-its-own -space-force/; David Reevely, "Canada Finalizing Plans for Its Version of US Space Force," The Logic, February 24, 2022, https://thelogic.co/news/canada-finalizing -plans-for-its-version-of-u-s-space-force/; Hanneke Weitering, "France Is Launch- ing a 'Space Force' With Weaponized Satellites," Space.com, August 2, 2019, https://www.space.com/france-military-space-force.html; NATO Secretary General Jens Stoltenberg, press conference following the meeting of the North Atlantic Council at the level of heads of state and/or government, North Atlan- tic Treaty Organization, December 4, 2019, https://www.nato.int/cps/en/natohq /opinions_171554.htm; Mari Yamaguchi, "Japan Launches New Unit to Boost Defense in Space," *Defense News*, May 18, 2020, https://www.defensenews .com/global/asia-pacific/2020/05/18/japan-launches-new-unit-to-boost-defense -in-space/.

52. Fred Kennedy, "It's Time to Equip the U.S. Space Force With the Ability to Project Force," *Forbes*, July 22, 2020, https://www.forbes.com/sites/fredkennedy /2020/07/22/its-time-to-equip-the-us-space-force-with-the-ability-to-project -force/.

53. Ibid.

54. Phillip Saunders, "Testimony Before the U.S.-China Economic and Security Review Commission," February 18, 2015, 8, http://www.uscc.gov/sites/default /files/Saunders_Testimony2.18.15.pdf.

55. Henry Sokolski, "China Waging War in Space: An After-Action Report," Occasional Paper No. 2104, Nonproliferation Policy Education Center, August 2021, https://npolicy.org/china-waging-war-in-space-an-after-action-report-occasional -paper-2104/; https://npolicy.org/wp-content/uploads/2021/08/2104-China -Space-Wargame-Report.pdf.

56. John P. Roth, Charles Brown, and John Raymond, "Department of the Air Force Posture Statement Fiscal Year 2022," https://docs.house.gov/meetings/AP/AP02/20210507/112533/HHRG-117-AP02-Bio-RaymondJ-20210507.pdf.

57. Ibid.

58. Brian Dunbar, "Gateway," NASA, April 4, 2022, https://www.nasa.gov/gateway/overview.

59. Peter Garretson, interview with Jean-Jacques Tortora, "A European Perspective," *Space Strategy*, podcast audio, January 31, 2022, https://anchor.fm/afpcspacepod/episodes/31--Jean-Jacques-Tortora-A-European-Perspective-e1dmsdd.

60. Ibid.; "Combined Space Operations Vision 2031," U.K. Ministry of Defense, February 22, 2022, https://www.gov.uk/government/publications/combined-space-operations-vision-2031.

61. Namrata Goswami and Peter Garretson, *Scramble for the Skies: The Great Power Competition to Control the Resources of Outer Space* (Lanham, MD: Lexington Books, 2020).

62. Park Si-soo, "US, India Agree to Cooperate on Space Situational Awareness," *SpaceNews*, April 12, 2022, https://spacenews.com/us-india-agree-to-cooperate-on-space-situational-awareness/.

63. Peter Garretson, "Sky's No Limit: Space-Based Solar Power, The Next Major Step in the Indo-US Strategic Partnership?" IDSA Occasional Paper No. 9, Institute for Defense Studies and Analyses, August 2010, https://spacejournal.ohio.edu/issue16/papers/OP_SkysNoLimit.pdf; Namrata Goswami and Peter Garretson, "The Rising Salience of 'NewSpace' in India: Prospects for U.S.-India Space Cooperation," *New Space* 10, no.1 (2022), https://www.liebertpub.com/doi/epdf/10.1089/space.2021.0038; Peter Garretson and Namrata Goswami, "Where Trump and Modi Should Take U.S.-India Space Cooperation," *National Interest*, https://nationalinterest.org/blog/buzz/where-trump-and-modi-should-take-us-india-space-cooperation-85896.

64. Yamaguchi, "Japan Launches New Unit to Boost Defense in Space."

65. "Research on the Space Solar Power Systems (SSPS)," Japan Aerospace Exploration Agency, https://www.kenkai.jaxa.jp/eng/research/ssps/ssps-index.html.

66. "Japan Tackles Clean Energy From Space," National Space Society, January 26, 2022, https://space.nss.org/japan-tackles-clean-energy-from-space/.

67. Stoltenberg, press conference following the meeting of the North Atlantic Council.

68. Weitering, "France Is Launching a 'Space Force' With Weaponized Satellites."

69. Jeff Foust, "Luxembourg Adopts Space Resources Law," *SpaceNews*, July 17, 2017, https://spacenews.com/luxembourg-adopts-space-resources-law/; Atossa Araxia Abrahamian, "How a Tax Haven Is Leading the Race to Privatise Space," *The Guardian*, September 15, 2017, https://www.theguardian.com/news/2017/sep/15/luxembourg-tax-haven-privatise-space.

70. Jonathan Porter, "UAE's Giant Leap Into Space," *National Geographic*, October 25, 2021, https://www.nationalgeographic.com/science/article/paid-content-uaes-giant-leap-into-space.

71. Rose Croshier, "Recommendations for US-Africa Space Cooperation and Development," Center for Global Development, March 31, 2022, https://

www.cgdev.org/publication/recommendations-us-africa-space-cooperation
-and-development; Rose Croshier, "Opportunities for US-Africa Space Cooperation and Development," Center for Global Development, CGD Policy Paper
256, March 31, 2022, https://www.cgdev.org/publication/opportunities
-us-africa-space-cooperation-and-development.

72. Jeffrey Hill, "Africa Unites Its Diverse and Rapidly Growing Space Industry," *Via Satellite*, October 16, 2020, https://interactive.satellitetoday.com/africa
-unites-its-diverse-and-rapidly-growing-space-industry/.

73. Croshier, "Recommendations for US-Africa Space Cooperation and
Development"; Croshier, "Opportunities for US-Africa Space Cooperation and
Development."

74. Shelli Brunswick, "Order and Progress–Brazil's Second Act in Space,"
SpaceNews, March 17, 2022, https://spacenews.com/op-ed-order-and-progress
-brazils-second-act-in-space/.

75. Tom Metcalfe, "Russia Could End Its Role in the International Space Station
by 2024, Say Experts," Live Science, April 6, 2022, https://www.livescience.com
/russia-iss-cooperation-end-2024.

76. Guo Zhaobing (郭照兵), *The Origin and Development of Sino-U.S. Space Cooperation, Obstacles, and Prospects* (Beijing: World Knowledge Press, 2020).

77. Leonard David, "Can the U.S. and China Cooperate in Space?" *Scientific
American*, August 2, 2021, https://www.scientificamerican.com/article/can-the
-u-s-and-china-cooperate-in-space/.

78. Dean Cheng, "U.S.-China Space Cooperation: More Costs Than Benefits,"
Heritage Foundation, October 30, 2009, https://www.heritage.org/space-policy
/report/us-china-space-cooperation-more-costs-benefits#; Larry M. Wortzel,
"China and the Battlefield in Space," Heritage Foundation, October 15, 2003,
https://www.heritage.org/asia/report/china-and-the-battlefield-space; "Cyber
Espionage and the Theft of U.S. Intellectual Property and Technology," Testimony
of Larry M. Wortzel Before the House of Representatives Committee on Energy and
Commerce Subcommittee on Oversight and Investigations, July 9, 2013 https://
www.uscc.gov/sites/default/files/Wortzel-OI-Cyber-Espionage-Intellectual
-Property-Theft-2013-7-9.pdf.

79. Dallas Boyd, Jeffrey Lewis, and Joshua Pollack, "Advanced Technology
Acquisition Strategies of the People's Republic of China," Defense Threat Reduc-
tion Agency Report No. ASCO 2010 021, September 2010, https://irp.fas.org
/agency/dod/dtra/strategies.pdf;MarkStokes,GabrielAlvarado,EmilyWeinstein,
and Ian Easton, "China's Space and Counterspace Capabilities and Activities,"
Prepared for the U.S.-China Economic and Security Review Commission, Project 2049 and Pointe Bello, March 30, 2020, https://www.uscc.gov/sites/default
/files/2020-05/China_Space_and_Counterspace_Activities.pdf.

CHAPTER 7

1. See Namrata Goswami, statement before the U.S.-China Economic and Security Review Commission hearing on "China in Space: A Strategic Competition?"
April 25, 2019, https://www.uscc.gov/sites/default/files/Namrata%20Goswami
%20USCC%2025%20April.pdf.

2. Stephen Chen, "China Speeds Up Moon Base Plan in Space Race Against the US," *South China Morning Post*, December 29, 2021, https://www.scmp.com/news/china/science/article/3161324/china-speeds-moon-base-plan-space-race-against-us.

3. Steven J. Butow, Thomas Cooley, Eric Felt, and Joel B. Mozer, "State of the Space Industrial Base 2020," Virtual Workshop, NewSpace New Mexico, May 4–7, 2020, https://assets.ctfassets.net/3nanhbfkr0pc/3TLlIb4Z2UZG7szZdyVFuf/bafb12c16a37ee673b1ba30e72935c07/State_of_the_Space_Industrial_Base_2020_Workshop_Report_July_2020_FINAL.pdf; Peter Garretson, Richard Harrison, and Lamont Colucci participated in the conference.

4. Steven J. Butow, Thomas Cooley, Eric Felt, and Joel B. Mozer, "State of the Space Industrial Base 2020," Department of Defense, July 2020, https://assets.ctfassets.net/3nanhbfkr0pc/3TLlIb4Z2UZG7szZdyVFuf/bafb12c16a37ee673b-1ba30e72935c07/State_of_the_Space_Industrial_Base_2020_Workshop_Report_July_2020_FINAL.pdf.

5. Ibid.

6. Brent Ziarnick, *Developing National Power in Space: A Theoretical Model* (Jefferson, NC: McFarland, 2015), 230.

7. Ibid.

8. Eamon Barrett, "China Wants to Be a Leading Space Power by 2045—And It's Getting There Fast," *Fortune*, May 30, 2021, https://fortune.com/2021/05/30/china-space-race-rocket-landing-mars-us/.

9. Cao Siqi, "China Mulls $10 Trillion Earth-Moon Economic Zone," *Global Times*, November 1, 2019, https://www.globaltimes.cn/page/201911/1168698.shtml.

10. Erin Rosenbaum and Donovan Russo, "China Plans a Solar Power Play in Space That NASA Abandoned Decades Ago," CNBC, March 17, 2019, https://www.cnbc.com/2019/03/15/china-plans-a-solar-power-play-in-space-that-nasa-abandoned-long-ago.html.

11. Stephen Chen, "China's Nuclear Spaceships Will Be 'Mining Asteroids and Flying Tourists' as It Aims to Overtake US in Space Race," *South China Morning Post*, November 17, 2017, https://sg.news.yahoo.com/china-nuclear-spaceships-mining-asteroids-133407429.html.

12. The State Council Information Office of the People's Republic of China, "Full Text: China's Space Program: A 2021 Perspective," China National Space Administration, January 28, 2022, http://english.www.gov.cn/archive/whitepaper/202201/28/content_WS61f35b3dc6d09c94e48a467a.html.

13. Andrew Jones, "China's Commercial Sector Finds Funding and Direction," *SpaceNews*, April 25, 2021, https://spacenews.com/chinas-commercial-sector-finds-funding-and-direction/.

14. "Military and Security Developments Involving the People's Republic of China 2021," Annual Report to Congress, Office of the Secretary of Defense, 2021, https://media.defense.gov/2021/Nov/03/2002885874/-1/-1/0/2021-CMPR-FINAL.PDF.

15. Mark Stokes, Gabriel Alvarado, Emily Weinstein, and Ian Easton, "China's Space and Counterspace Capabilities and Activities," Prepared for the U.S.-China Economic and Security Review Commission, March 30, 2020, https://www.uscc.gov/research/chinas-space-and-counterspace-activities.

16. See, for example, Butow et al., "The State of the Space Industrial Base 2020"; J. Olson, S. Butow, E. Felt, T. Cooley, and J. Mozer, "State of the Space Industrial Base 2021: Infrastructure & Services for Economic Growth & National Security," Department of Defense, November 2021, https://assets.ctfassets.net/3na nhbfkr0pc/43TeQTAmdYrym5DTDrhjd3/a37eb4fac2bf9add1ab9f71299392043 /Space_Industrial_Base_Workshop_2021_Summary_Report_-_Final_15 _Nov_2021c.pdf; Bruce Cahan and Mir Sadat, "US Space Policies for the New Space Age: Competing on the Final Economic Frontier," based on *Proceedings from the State of Space Industrial Base 2020*, January 6, 2021, https://www.politico .com/f/?id=00000177-9349-d713-a777-d7cfce4b0000.

17. Michael Sheetz, "NASA Awards Blue Origin, Northrop Grumman and Nanoracks With Contracts to Build Private Space Stations," CNBC, December 2, 2021, https://www.cnbc.com/2021/12/02/nasa-private-space-station-contracts -blue-origin-nanoracks-northrop.html.

18. "The Fight Is on at Space Systems Command," Space Systems Command, March 4, 2022, https://www.ssc.spaceforce.mil/News/Article-Display /Article/2955844/the-fight-is-on-at-space-systems-command.

19. Butow et al., "The State of the Space Industrial Base 2020," 3, 15; Olson et al., "State of the Space Industrial Base 2021," 17, 34, 76, and B-2.

20. Space Commodities Exchange, http://www.spacecommoditiesexchange. com/product-listings/.

21. "Start-Up Space: Update on Investment in Commercial Space Ventures," BryceTech, 2022, https://brycetech.com/reports/report-documents/Bryce_Start _Up_Space_2022.pdf.

22. "Space Corporation Act of 2021," Foundation for the Future, https://www .f4f.space/_files/ugd/94f02a_a5e19d9f3a5b4becab138d9a747afabe.pdf.

23. Scott Phillips, "The New Frontier Playbook," 2017, https://newfrontierplay-book.com/chapter-4-l1-strategic-materials-reserve/.

24. Joshua Carlson, *Spacepower Ascendant: Space Development Theory and a New Space*, independently published, June 27, 2020, https://www.amazon .com/Spacepower-Ascendant-Development-Theory-Strategy-ebook/dp/B08B Y52LFN.

25. Joshua Carlson, personal communication, April 2022.

26. BeiDou Navigation Satellite System, http://en.beidou.gov.cn/.

27. "2022 Challenges to Security in Space," U.S. Defense Intelligence Agency, iii and 1, https://www.dia.mil/Portals/110/Documents/News/Military_Power _Publications/Challenges_Security_Space_2022.pdf.

28. Andrew Jones, "China Is Developing Plans for a 13,000-Satellite Megaconstellation," *SpaceNews*, April 21, 2022, https://spacenews.com/china-is-developing -plans-for-a-13000-satellite-communications-megaconstellation/.

29. Yen Nee Lee, "China Races to Rival the U.S. With Its Own GPS System—But One Analyst Says It Won't Overtake the U.S. Yet," CNBC, May 31, 2021, https:// www.cnbc.com/2021/06/01/tech-war-chinas-beidou-gains-market-share -challenges-us-gps.html.

30. Andrew Jones, "China Establishes Company to Build Satellite Broadband Megaconstellation," *SpaceNews*, May 26, 2021, https://spacenews.com /china-establishes-company-to-build-satellite-broadband-megaconstellation/.

31. "China Cyber Threat Overview and Advisories," U.S. Cybersecurity & Infrastructure Security Agency, https://www.cisa.gov/uscert/china.

32. Karen Kwon, "Space-Based Quantum Communications," *Scientific American*, June 25, 2020, https://www.scientificamerican.com/article/china-reaches-new-milestone-in-space-based-quantum-communications/.

33. Luyuan Xu, "How China's Lunar Relay Satellite Arrived in Its Final Orbit," Planetary Society, June 15, 2018, https://www.planetary.org/articles/20180615-queqiao-orbit-explainer.

34. Stephen Chen, "Chinese Satellite Set to Join 'Radio Telescope' Covering Area 30 Times the Size of Earth," *South China Morning Post*, March 31, 2022, https://www.scmp.com/news/china/science/article/3172634/chinese-satellite-set-create-radio-telescope-30-times-size-earth.

35. Neel V. Patel, "China's Surging Private Space Industry Is Out to Challenge the US," *MIT Technology Review*, January 21, 2021, https://www.technologyreview.com/2021/01/21/1016513/china-private-commercial-space-industry-dominance/.

36. Alan C. O'Connor et al., "Economic Benefits of the Global Positioning System (GPS)," Final Report Prepared for the National Institute of Standards and Technology, RTI International, June 2019, https://www.rti.org/sites/default/files/gps_finalreport.pdf.

37. "GPS Modernization DOD Continuing to Develop New Jam-Resistant Capability, but Widespread Use Remains Years Away," Report to Congressional Committees, GAO-21-145 (Washington, DC: U.S. Government Accountability Office, January 2021), https://www.gao.gov/assets/gao-21-145.pdf.

38. Jenna Mukuno and Krissy Eliot, "Our Constellation," Planet, https://www.planet.com/company/.

39. Mike Nichols, "Forecast to Industry," Commercial Satellite Communications Office, U.S. Space Force, October 29, 2021, https://www.disa.mil/-/media/Files/DISA/News/Events/Forecast-to-Industry---2021/01---US-Space-Force-Commercial-Satellite-Communications-Office---Nichols.ashx; Nathan Strout, "Space Force Will Set Up One Office for Commercial Services, Including SATCOM and Satellite Imagery," C4ISR, June 2, 2021, https://www.c4isrnet.com/battlefield-tech/space/2021/06/02/space-force-will-set-up-one-office-for-commercial-space-services-including-satcom-and-satellite-imagery/.

40. Avery Thompson, "China Wants a Nuclear Space Shuttle by 2040," *Popular Mechanics*, November 16, 2017, https://www.popularmechanics.com/space/rockets/a13788331/chinas-future-space-plans/.

41. Eric Berger, "China's State Rocket Company Unveils Rendering of a Starship Look-Alike," *Ars Technica*, April 26, 2021, https://arstechnica.com/science/2021/04/chinas-state-rocket-company-unveils-rendering-of-a-starship-look-alike/.

42. Chen, "China's Nuclear Spaceships Will Be 'Mining Asteroids and Flying Tourists.'"

43. Andrew Jones, "China's Shenzhou 13 Astronauts Manually Fly Cargo Ship for Emergency Docking Test," *Space.com*, January 13, 2022, https://www.space.com/china-shenzhou-13-emergency-docking-test.

44. See Goswami, "China in Space: A Strategic Competition?"

45. Brian Dunbar, "Commercial Orbital Transportation Services (COTS)," Commercial Space Economy, NASA, August 3, 2017, https://www.nasa.gov/commercial-orbital-transportation-services-cots.

46. Brian Dunbar, "Commercial Crew Program Overview," Commercial Crew Press Kit, NASA, July 27, 2021, https://www.nasa.gov/content/commercial-crew-program-overview.

47. Brian Dunbar, "Commercial Lunar Payload Services Overview," Commercial Lunar Payload Services, NASA, June 24, 2021, https://www.nasa.gov/content/commercial-lunar-payload-services-overview.

48. Brian Dunbar, "In-Space Manufacturing," Off-Earth Manufacturing, NASA, May 22, 2019, https://www.nasa.gov/oem/inspacemanufacturing.

49. Brian Dunbar, "NASA Innovative Advanced Concepts (NIAC)," STMD: NIAC, NASA, April 8, 2021, https://www.nasa.gov/directorates/spacetech/niac/index.html.

50. Brian Dunbar, "NASA's Centennial Challenges Overview," STMD: Centennial Challenges, NASA, June 25, 2021, https://www.nasa.gov/directorates/spacetech/centennial_challenges/overview.html.

51. Sandra Erwin, "Space Force Launches 'Orbital Prime' Program to Spur Market for On-Orbit Services," *SpaceNews*, November 4, 2021, https://spacenews.com/space-force-launches-orbital-prime-program-to-spur-market-for-on-orbit-services/.

52. Nichols Martin, "DIU Seeks Vehicles, Fuel Depots for Space Logistics," *Executive Biz*, February 6, 2020, https://blog.executivebiz.com/2020/02/diu-seeks-vehicles-fuel-depots-for-space-logistics/.

53. Jeff Foust, "Three Companies Studying 'Orbital Outpost' Space Station Concepts for Defense Department," *SpaceNews*, July 19, 2020, https://spacenews.com/three-companies-studying-orbital-outpost-space-station-concepts-for-defense-department/.

54. Nathan Greiner and Tabitha Dodson, "Demonstration Rocket for Agile Cislunar Operations (DRACO)," Defense Advanced Research Project Agency, https://www.darpa.mil/program/demonstration-rocket-for-agile-Cislunar-operations.

55. "Orbital Construction: DARPA Pursues Plan for Robust Manufacturing in Space," Defense Advanced Research Project Agency, February 2, 2021, https://www.darpa.mil/news-events/2021-02-05.

56. Sean Potter, "NASA Publishes Artemis Plan to Land First Woman, Next Man on Moon in 2024," NASA, January 4, 2021, https://www.nasa.gov/press-release/nasa-publishes-artemis-plan-to-land-first-woman-next-man-on-moon-in-2024.

57. "Nasa Artemis Moon Mission Delayed to 2025," BBC, November 11, 2021, https://www.bbc.co.uk/newsround/59232938.

58. "Artemis: NASA's Crewed Mission Delayed Till 2026," *WION News*, March 4, 2022, https://www.wionews.com/science/artemis-nasas-crewed-mission-delayed-till-2026-458714.

59. Jeff Foust, "Pence Calls for Human Return to the Moon by 2024," *SpaceNews*, March 26, 2019, https://spacenews.com/pence-calls-for-human-return-to-the-moon-by-2024/.

60. "Space Corporation Act of 2021," Foundation for the Future, https://www.f4f.space/_files/ugd/94f02a_a5e19d9f3a5b4becab138d9a747afabe.pdf.

61. Armen Papazian, *The Space Value of Money: Rethinking Finance Beyond Risk and Time* (London: Palgrave Macmillan, 2022), https://link.springer.com /book/10.1057/978-1-137-59489-1.

62. Cahan and Sadat, "US Space Policies for the New Space Age."

63. Scott Phillips, "The New Frontier Playbook," 2017, https://newfrontierplay-book.com/chapter-4-l1-strategic-materials-reserve/.

64. Dennis Wingo, "The Early Space Age, the Path Not Taken Then, but Now? (Part 1)," February 16, 2015, https://denniswingo.wordpress.com/2015/02/16 /the-early-space-age-the-path-not-taken-then-but-now/.

65. Michael Le Page, "Crew of Mock Lunar 'Biosphere' Grew Food and Made Oxygen for 200 Days," *New Scientist*, March 2, 2021, https://www.newscientist .com/article/2268836-crew-of-mock-lunar-biosphere-grew-food-and-made-oxygen-for-200-days/.

66. Andrew Jones, "Chinese Crewed Moon Landing Possible by 2030, Says Senior Space Figure," *SpaceNews*, November 15, 2021, https://spacenews.com /chinese-crewed-moon-landing-possible-by-2030-says-senior-space-figure/; Ryan Woo and Liangping Gao, "China Plans Its First Crewed Mission to Mars in 2033," Reuters, June 24, 2021, https://www.reuters.com/business/aerospace-defense /china-plans-its-first-crewed-mission-mars-2033-2021-06-24/.

67. "Astronauts Return to Earth After China's Longest Space Mission," BBC News, September 17, 2021, https://www.bbc.com/news/world-asia-china-58554332.

68. "Space Tourism—Global Market Trajectory & Analytics," Global Industry Analysts Report, April 2021, https://www.researchandmarkets.com /reports/5141552/space-tourism-global-market-trajectory-and.

69. Elon Musk, "Making Humans a Multi-Planetary Species," *New Space* 5, no. 2 (2017), https://www.liebertpub.com/doi/10.1089/space.2017.29009.emu.

70. Catherine Clifford, "Jeff Bezos Dreams of a World With a Trillion People Living in Space," CNBC, May 1, 2018, https://www.cnbc.com/2018/05/01 /jeff-bezos-dreams-of-a-world-with-a-trillion-people-living-in-space.html.

71. Eric Mack, "Elon Musk Has New Estimate for When Humans Might First Step on Mars," CNET, March 17, 2022, https://www.cnet.com/science/space /elon-musk-has-new-estimate-for-when-humans-might-first-step-on-mars/.

72. "The Northwest Ordinance For Space," Citizens in Space, http://www .citizensinspace.org/2014/04/the-northwest-ordinance-for-space/; Rep. George Brown (D-Calif.) introduced the "Space Settlement Act of 1988," H.R. 4218 (115), 7; H.R. 4218, NSS, https://space.nss.org/media/Space-Settlement-Act-Of-1988 .pdf; Public Law 100-685, November 17, 1988, GovInfo, https://www.govinfo .gov/content/pkg/STATUTE-102/pdf/STATUTE-102-Pg4083.pdf; National Aeronautics and Space Act of 1958, Public Law 85-568, Stat. 426 (1958); Proposed "The Space Development and Settlement Act of 2019," Alliance for Space Development, http://allianceforspacedevelopment.org/wp-content/uploads/2019/01/Space -Development-and-Settlement-Act-of-2019.pdf.

73. Brian Dunbar, "About the Space Station Solar Arrays," International Space Station, NASA, August 3, 2017, https://www.nasa.gov/mission_pages/station /structure/elements/solar_arrays-about.html.

74. Stephen Chen, "China's Space Programme Will Go Nuclear to Power Future Missions to the Moon and Mars," *South China Morning Post*, November

24, 2021, https://www.scmp.com/news/china/science/article/3157213/chinas -space-programme-will-go-nuclear-power-future-missions.

75. "Full Text: China's Space Program: A 2021 Perspective," White Paper, The State Council Information Office of the People's Republic of China, January 28, 2022, http://english.www.gov.cn/archive/whitepaper/202201/28/conten t_WS61f35b3dc6d09c94e48a467a.html.

76. Gao Ji, Hou Xinbin, and Wang Li, "Solar Power Satellites Research in China," *Online Journal of Space Communication* 16 (Winter 2010), https://spacejournal.ohio .edu/issue16/ji.html.

77. "China Plans to Exploit $10 Trillion Earth-Moon Economic Zone," American Security Council Foundation, December 2, 2020, https://www.ascf.us/news /china-plans-to-exploit-10-trillion-earthmoon-economic-zone/.

78. David Whitehouse, "China Denies Manned Moon Mission Plans," BBC News, May 21, 2002, http://news.bbc.co.uk/2/hi/science/nature/2000506. stm.

79. Dave Makichuk, "Helium-3: The Secret 'Mining War' in Space," *Asia Times*, November 6, 2021, https://asiatimes.com/2021/11/helium-3-the-secret -mining-war-in-space/.

80. Didi Tang, "China Embarks on Space Race for Solar Power," *The Times (UK)*, August 17, 2021, https://www.thetimes.co.uk/article/china-embarks -on-space-race-for-solar-power-2hvkdhlcx.

81. Brian Dunbar, "Kilopower," Space Technology Mission Directorate, NASA, October 18, 2021, https://www.nasa.gov/directorates/spacetech/kilopower.

82. "NASA Seeks Proposals for Lunar Reactor," *World Nuclear News*, November 22, 2021, https://www.world-nuclear-news.org/Articles/NASA-seeks-proposals -for-lunar-reactor; Brian Dunbar, "Fission Surface Power," Space Technology Mission Directorate, NASA, March 14, 2021, https://www.nasa.gov/mission_pages /tdm/fission-surface-power/index.html.

83. Greiner and Dodson, "Demonstration Rocket for Agile Cislunar Operations (DRACO)."

84. "FY 22 Defense Appropriations Bill," https://docs.house.gov/billsthisweek /20220307/BILLS-117RCP35-JES-DIVISION-C_Part2.pdf.

85. "Space Solar Power Incremental Demonstrations and Research Project (SSPIDR)," Space Power Beaming, Air Force Research Laboratory, https:// afresearchlab.com/technology/space-power-beaming/; "FY 22 Defense Appropriations Bill"; "RDT&E Programs," Fiscal Year 2023 Department of Defense Budget, Program Element Number 1106857SF, https://comptroller.defense.gov /Portals/45/Documents/defbudget/FY2023/FY2023_r1.pdf.

86. Namrata Goswami and Peter Garretson, *Scramble for the Skies: The Great Power Competition to Control the Resources of Outer Space* (Lanham, MD: Lexington Books, 2020).

87. Andrew Jones, "China's Super Heavy Rocket to Construct Space-Based Solar Power Station," *SpaceNews*, June 28, 2021, https://spacenews.com /chinas-super-heavy-rocket-to-construct-space-based-solar-power-station/; Jeff Brown, "China Is Building a Solar Power Plant in Space," Brownstone Research, July 7, 2021, https://www.brownstoneresearch.com/bleeding-edge /china-is-building-a-solar-power-plant-in-space/.

88. "Exploiting Earth-Moon Space: China's Ambition After Space Station," *Xinhua*, March 18, 2016, https://www.chinadaily.com.cn/china/2016-03/08/content_23775949.htm.

89. Ming Li, "Using Resources on Asteroid for Manufacturing Of SSPS," China Academy of Space Technology (CAST), China-International Astronautical Congress-September 26, Guadalajara, Mexico, https://www.youtube.com/watch?v=6kum9VbVmN8.

90. Andrew Jones, "China Moon Rock Studies Include Fusion Energy Analysis, Volcanic History," Space.com, September 17, 2021, https://www.space.com/china-moon-rocks-as-fusion-energy-source; Adam Xu, "China Joins Race to Mine Moon for Resources," VOANews.com, December 2, 2020, https://www.voanews.com/a/science-health_china-joins-race-mine-moon-resources/6199085.html.

91. Tia Vialva, "China National Space Administration to Establish 3D Printed Houses on the Moon," 3D Printing Industry, January 15, 2021, https://3dprintingindustry.com/news/china-national-space-administration-to-establish-3d-printed-houses-on-the-moon-147133/; Carlota V, "China Reveals Plans to Build 3D Printed Base on Moon," *3D Printing News*, January 18, 2019, https://www.3dnatives.com/en/china-3d-printed-base-moon-180120195/.

92. Andrew Jones, "China Launches Space Mining Test Spacecraft on Commercial Rideshare Mission," *SpaceNews*, April 27, 2021, https://spacenews.com/china-launches-space-mining-test-spacecraft-on-commercial-rideshare-mission/.

93. Ryan Whitwam, "Chinese Scientists Want to Capture a Small Asteroid and Land It on Earth," *Extreme Tech*, July 30, 2018, https://www.extremetech.com/extreme/274388-chinese-scientists-want-to-capture-a-small-asteroid-and-land-it-on-earth; Nicole Arce, "China Plans to Capture an Asteroid and Bring It Down to Earth," *Tech Times*, July 26, 2018, https://www.techtimes.com/articles/232692/20180726/china-plans-to-capture-an-asteroid-and-bring-it-down-to-earth.htm; "China Focus: Capture an Asteroid, Bring It Back to Earth?" *Xinhua*, July 24, 2018, https://www.chinadaily.com.cn/a/201807/24/WS5b-568b56a31031a351e8fbbd.html; "China Plans to Capture Asteroid and Bring It Back to Earth," *TomoNews*, August 3, 2018, https://youtu.be/M8hyzq2h6Ng.

94. Government of Luxembourg Ministry of Economy, "Luxembourg Cooperates With China in the Exploration and Use of Outer Space for Peaceful Purpose, Including in the Utilization of Space Resources," Space Resources.lu, January 18, 2018, https://spacelu.com/wp-content/uploads/2019/06/10-2018-01-17-press-release-cooperation-china-luxembourg.pdf.

95. Andrew Jones, "China Has Big Plans for Its New Tiandu Space Exploration Laboratory," Space.com, March 31, 2022, https://www.space.com/china-tiandu-deep-space-exploration-laboratory.

96. Dunbar, "In-Space Manufacturing."

97. "Orbital Construction: DARPA Pursues Plan for Robust Manufacturing in Space."

98. The White House, "In-Space Servicing, Assembly, and Manufacturing National Strategy," Office of Science and Technology Policy, April 2022, https://www.whitehouse.gov/wp-content/uploads/2022/04/04-2022-ISAM-National-Strategy-Final.pdf.

99. "FY 2023 NASA Budget Request," https://www.nasa.gov/sites/default/files/atoms/files/fy23_nasa_budget_request_summary.pdf.

100. Dennis Wingo, "The Lunar Industrial Facility and Orbital Shipyards; How to Get There," July 31, 2017, https://denniswingo.wordpress.com/2017/07/31/the-lunar-industrial-facility-and-orbital-shipyards-how-to-get-there/.

101. H.R.2262—U.S. Commercial Space Launch Competitiveness Act, Public Law 114-90, 114th Congress, November 25, 2015, https://www.congress.gov/bill/114th-congress/house-bill/2262/text?overview=closed.

102. For example, NASA buys Lunar regolith. See Brian Dunbar, "NASA Selects Companies to Collect Lunar Resources for Artemis Demonstrations," NASA, January 4, 2020, https://www.nasa.gov/press-release/nasa-selects-companies-to-collect-lunar-resources-for-artemis-demonstrations.

103. Subject matter experts include CEOs of private-sector space companies, leaders in former top U.S. government space positions, high-ranking military members, and leading think tank and academic professionals.

104. Nicholas Eftimiades, "Small Satellites: The Implications for National Security," Atlantic Council, May 2022, https://www.atlanticcouncil.org/wp-content/uploads/2022/05/Small_satellites-Implications_for_national_security.pdf.

105. The quote from Rick Tumlinson was provided via email to Peter Garretson and Richard Harrison on June 30, 2022.

CHAPTER 8

1. "The Future of Space 2060 and Implications for U.S. Strategy: Report on the Space Futures Workshop," Air Force Space Command, September 5, 2019, https://www.afspc.af.mil/Portals/3/documents/Future%20of%20Space%202060%20v2%20(5%20Sep).pdf. A graphically updated version of the report was released in October 2019 and is available at https://www.afspc.af.mil/Portals/3/The%20Future%20of%20Space%202060%20-%203Oct19.pdf. These sentiments are echoed further in The White House, "A New Era for Deep Space Exploration and Development," National Space Council, July 23, 2020, https://spp.fas.org/eprint/new-era-2020.pdf; Steven J. Butow, Thomas Cooley, Eric Felt, and Joel B. Mozer, "The State of the Space Industrial Base 2020," U.S. Department of Defense, July 2020, https://www.newspacenm.org/wp-content/uploads/2020/07/State-of-the-Space-Industrial-Base-2020_A-Time-for-Action-to-Sustain-US-Econ.-Mil-Leadership-in-Space.pdf; and Namrata Goswami and Peter Garretson, *Scramble for the Skies: The Great Power Competition to Control the Resources of Outer Space* (Lanham, MD: Lexington Press, 2020).

2. U.S.-China Economic and Security Review Commission, "2019 Report to Congress," November 2019, https://www.uscc.gov/sites/default/files/2019-11/Chapter%204%20Section%203%20-%20China%E2%80%99s%20Ambitions%20in%20Space%20-%20Contesting%20the%20Final%20Frontier.pdf.

3. Ibid.

4. "About NASA," National Aeronautics and Space Administration, December 11, 2019, https://www.nasa.gov/about/index.html.

5. See, for example, the debate on NASA authorization: Jeff Foust, "House Subcommittee Advances NASA Authorization Bill," *SpaceNews*, January 30, 2020, https://spacenews.com/house-subcommittee-advances-nasa-authorization-bill/.

6. Courtney Johnson, "How Americans See the Future of Space Exploration, 50 Years After the First Moon Landing," Pew Research Center, July 17, 2019, https://

www.pewresearch.org/fact-tank/2019/07/17/how-americans-see-the-future-of
-space-exploration-50-years-after-the-first-Moon-landing/.

7. Lori Garver, "Forget New Crewed Missions in Space. NASA Should Focus on
Saving Earth," *Washington Post*, July 18, 2019, https://www.washingtonpost.com
/opinions/forget-new-manned-missions-in-space-nasa-should-focus-on-saving
-earth/2019/07/18/79e55eb8-a995-11e9-9214-246e594de5d5_story.html.

8. "Why Do We Send Robots to Space?" NASA Science Space Place, July 27,
2020, https://spaceplace.nasa.gov/space-robots/en/.

9. Nola Taylor Redd, "Planetary Protection: Contamination Debate Still Sim-
mers," Space.com, May 8, 2017, https://www.space.com/36708-planetary
-protection-astrobiology-nasa-missions.html. Sending robots does not ensure
there is no contamination, as evidenced by a crashed Israeli lander on the Moon.
See Daniel Oberhaus, "A Crashed Israeli Lunar Lander Spilled Tardigrades on
the Moon," *Wired*, August 5, 2019, https://www.wired.com/story/a-crashed
-israeli-lunar-lander-spilled-tardigrades-on-the-Moon/.

10. Robert Zubrin, "Wokeists Assault Space Exploration," *National Review*,
November 14, 2020, https://www.nationalreview.com/2020/11/wokeists-assault
-space-exploration/.

11. James Oberg, "Planetary Defense, Asteroid Deflection & the Future of Human
Intervention in the Earth's Biosphere," Futures Focus Day Symposium sponsored
by Commander-in-Chief, US Space Command Colorado Springs, Colorado, July
23, 1998, http://defendgaia.org/bobk/oberg.html.

12. Thomas Matula and Karen Loveland, "Public Attitudes Toward Different
Space Goals: Building Public Support for the Vision for Space Exploration (VSE),"
10th Biennial International Conference on Engineering, Construction, and Operations in Challenging Environments and Second NASA/ARO/ASCE Workshop on
Granular Materials in Lunar and Martian Exploration, March 5–8, 2006, https://
ascelibrary.org/doi/10.1061/40830%28188%2958; see also "50% Favor Cutting
Back on Space," Billion Year Plan, January 17, 2010, http://billionyearplan.blogspot
.com/2010/01/50-favor-cutting-back-on-space.html.

13. Ioana Cozmuta and Adam Routh, "From 'Flags and Footprints' to Having a
Routine Presence in Space," *The Hill*, January 1, 2018, https://thehill.com/opinion
/technology/366950-from-flags-and-footprints-to-the-us-having-a-routine
-presence-in-space.

14. "Elon Musk's SpaceX Sues Government to Protest Military Launch Monopoly," NBC News, April 25, 2014, https://www.nbcnews.com/science/space/elon
-musks-spacex-sues-government-protest-military-launch-monopoly-n89926;
Christian Brose, *The Kill Chain: Defending America in the Future of High Tech Warfare*
(New York: Hachette Books, 2020), 72.

15. Brose, *The Kill Chain*, 72.

16. Rachel Olney, "The Rift Between Silicon Valley and the Pentagon Is Economic, Not Moral," War on the Rocks, January 28, 2019, https://warontherocks
.com/2019/01/the-rift-between-silicon-valley-and-the-pentagon-is-economic
-not-moral/.

17. Frank Hoffman, "The Hypocrisy of the Techno-Moralists in the Coming
Age of Autonomy," War on the Rocks, March 6, 2019, https://warontherocks.
com/2019/03/the-hypocrisy-of-the-techno-moralists-in-the-coming-age-of
-autonomy/; David Ignatius, "Can the Pentagon Build a Bridge to the Tech

Community?" January 24, 2019, https://www.washingtonpost.com/opinions /can-the-pentagon-build-a-bridge-to-the-tech-community/2019/01 /24/39c0e3b2-2019-11e9-9145-3f74070bbdb9_story.html.

18. Jenny Bavisotto, "China's Military-Civil Fusion Strategy Poses a Risk to National Security," DipNote: Military and Security, U.S. Department of State Bureau of International Security and Nonproliferation, January 30, 2020, https://2017-2021.state.gov/chinas-military-civil-fusion-strategy-poses-a-risk-to -national-security/index.html.

19. Courtney Stadd, "Industry Insight: Space Jobs of the Future," *Space Report 2019, Quarterly 3*, https://www.thespacereport.org/uncategorized/industry -insight-space-jobs-of-the-future/; Domingo Angeles and Dennis Vilorio, "Space Careers: A Universe of Options," U.S. Bureau of Labor Statistics, November 2016, https://www.bls.gov/careeroutlook/2016/article/careers-in-space.htm.

20. Butow et al., "The State of the Space Industrial Base 2020."

21. On the need to develop and retain space professionals, see, for example, The White House, "National Space Policy of the United States of America," June 28, 2010, 8, https://obamawhitehouse.archives.gov/sites/default/files/national _space_policy_6-28-10.pdf.

22. Michael T. Nietzel, "U.S. Universities Fall Further Behind China in Pro-duction of STEM PhDs," *Forbes*, August 7, 2021, https://www.forbes.com/sites /michaeltnietzel/2021/08/07/us-universities-fall-behind-china-in-production -of-stem-phds/?sh=18775ee74606.

23. "Scientific Brain Drain: Quantifying the Decline of the Federal Scientific Work-force," A Majority Staff Report Prepared for Members of the U.S. House of Rep-resentatives Committee on Science, Space, and Technology, March 2021, https:// science.house.gov/imo/media/doc/2021-3%20EMBARGOED%20Scientific%20 Brain%20Drain%20Majority%20STAFF%20REPORT%20w%20cover%20page.pdf.

24. Alfred Thayer Mahan, *The Influence of Sea Power on History, 1660–1783* (Bos-ton: Little, Brown and Company, 1890).

25. Dustin Grant and Matthew Neil, "The Case for Space: A Legislative Frame-work for an Independent United States Space Force," Air Command and Staff Col-lege, February 2020, https://www.airuniversity.af.edu/Portals/10/AUPress/Papers /WF_73_GRANTNEIL_THE_CASE_FOR_SPACE_A_LEGISLATIVE_FRAME WORK_FOR_AN_INDEPENDENT_UNITED_STATES_SPACE_FORCE.PDF.

26. U.S. Department of Defense, "Directive 5100.01," September 17, 2020, https://www.esd.whs.mil/Portals/54/Documents/DD/issuances/dodd/510001p .pdf.

27. Armen Papazian, *The Space Value of Money: Rethinking Finance Beyond Risk and Time* (London: Palgrave Macmillan, 2022).

28. "Space Corporation Act of 2021," Foundation for the Future, https://www .f4f.space/_files/ugd/94f02a_a5e19d9f3a5b4becab138d9a747afabe.pdf.

29. Bruce B. Cahan, R. Pittman, Sarah Cooper, and J. Cumbers, "Space Com-modities Futures Trading Exchange: Adapting Terrestrial Market Mechanisms to Grow a Sustainable Space Economy," *New Space*, September 2018, https://www .liebertpub.com/doi/10.1089/space.2017.0047.

30. Scott Phillips, "Chapter 4: L1 Strategic Materials Reserve," New Frontier Playbook, https://newfrontierplaybook.com/chapter-4-l1-strategic-materials -reserve/.

31. David Sobyra and Melissa Jacobs, "Security Cooperation Billing Handbook," Defense Security Cooperation University (DSCU) and Defense Finance and Accounting Service (DFAS) Security Cooperation Accounting (SCA) Edition 14, January 2021, https://www.dscu.edu/documents/publications/security-cooperation-billing-handbook.pdf.

32. Amy Burke, Abigail Okrent, and Katherine Hale, "The State of U.S. Science and Engineering 2022," National Science Foundation, January 18, 2022, https://ncses.nsf.gov/pubs/nsb20221/executive-summary.

33. Marcia Smith, "Biden Administration Embraces Office of Space Commerce in FY 2023 Budget," SpacePolicyOnline.com, March 29, 2022, https://spacepolicyonline.com/news/biden-administration-embraces-office-of-space-commerce-in-fy2023-budget/.

34. "Decadal Science Strategy Surveys: Report of a Workshop," National Resource Council, 2007, https://nap.nationalacademies.org/read/11894/chapter/3#7.

35. "The Artemis Accords," NASA, https://www.nasa.gov/specials/artemis-accords/index.html.

36. "Frequently Asked Questions," U.S. Transportation Command (USTRANSCOM), https://www.ustranscom.mil/dbw/faq.cfm.

37. "Space Systems Command: Securing Secure Comms," *Milsat Magazine*, February 20222, http://milsatmagazine.com/story.php?number=1585555271.

38. Loren Blinde, "DIU Seeks Responsive Space Launch Services," *Intelligence Community News*, August 3, 2020, https://intelligencecommunitynews.com/diu-seeks-responsive-space-launch-services/.

39. Butow et al., "The State of the Space Industrial Base 2020"; J. Olson, S. Butow, E. Felt, T. Cooley, and J. Mozer, "State of the Space Industrial Base 2021: Infrastructure & Services for Economic Growth & National Security," U.S. Department of Defense, 17, 34, 76, B-2, November 2021, https://assets.ctfassets.net/3nanhbfkr0pc/43TeQTAmdYrym5DTDrhjd3/a37eb4fac2bf9add1ab9f71299392043/Space_Industrial_Base_Workshop_2021_Summary_Report_-_Final_15_Nov_2021c.pdf.

40. "Civil Reserve Airfleet," U.S. Department of Transportation, https://www.transportation.gov/mission/administrations/intelligence-security-emergency-response/civil-reserve-airfleet-allocations; Christopher Bolkcom, "Civil Reserve Air Fleet (CRAF)," CRS Report for Congress, RL 33692, October 18, 2006, https://sgp.fas.org/crs/weapons/RL33692.pdf.

41. "Civil Reserve Air Fleet (CRAF) Aircraft," Global Security.org, https://www.globalsecurity.org/military/library/policy/army/fm/55-9/ch3.htm.

42. Peter Garretson, "Establish Space National Guard now," *The Hill*, June 7, 2022, https://thehill.com/opinion/national-security/3514195-establish-the-space-national-guard-now/.

43. The White House, "In-Space Servicing, Assembly, and Manufacturing National Strategy," National Science and Technology Council, April 2022, https://www.whitehouse.gov/wp-content/uploads/2022/04/04-2022-ISAM-National-Strategy-Final.pdf.

44. "Defense Production Act, Title III Program," https://www.ndia.org/-/media/sites/press-media/dpa-title-iii-overview.ashx.

45. Ibid.; Matthew Seaford, "Title III of the Defense Production Act," U.S. Department of Energy, https://www.energy.gov/sites/prod/files/2014/03/f14/2_seaford_roundtable.pdf.

46. Peter Garretson, "Clarifying the Planetary Defense Mission," *Defense Technology Program Brief* No. 24, June 2021, https://www.afpc.org/uploads/documents/Defense_Technology_Briefing_-_Issue_24.pdf.

47. "USSPACECOM Deputy Presents on Planetary Defense at 37th Annual Space Symposium, United States Space Command," April 6, 2022, https://www.dvidshub.net/image/7137185/usspacecom-deputy-presents-planetary-defense-37th-annual-space-symposium; Ramin Skibba, "Space Command's Lt. Gen John Shaw Says Space Is 'Under Threat,'" *Wired*, April 15, 2022, https://www.wired.com/story/space-commands-lt-gen-john-shaw-on-the-future-of-space-security/.

48. Butow et al., "The State of the Space Industrial Base 2020"; Olson et al., "State of the Space Industrial Base 2021."

49. "Joint Interagency Coordination Group (JIACG) A Prototyping Effort," U.S. Joint Forces Command Fact Sheet, January 2005, https://smallwarsjournal.com/documents/jiacgfactsheet.pdf.

50. Johnathan Ward, *China's Vision of Victory* (Arlington, VA: Atlas Publishing and Media Company, 2019).

51. James Black, Linda Slapakova, and Kevin Martin, "Future Uses of Space Out to 2050: Emerging Threats and Opportunities for the UK National Space Strategy," RAND Corporation, 2022, https://www.rand.org/content/dam/rand/pubs/research_reports/RRA600/RRA609-1/RAND_RRA609-1.pdf.

Contributors

Richard M. Harrison is the vice president of operations and director of defense technology programs at the American Foreign Policy Council (AFPC). He currently serves as managing editor of AFPC's *Defense Dossier* e-journal and as editor of the *Defense Technology Monitor* e-bulletin. He also directs a briefing series on Capitol Hill to educate congressional staff on defense technology issues affecting U.S. national security. Mr. Harrison coedited a critically acclaimed volume on cybersecurity, titled *Cyber Insecurity: Navigating the Perils of the Next Information Age* (Rowman & Littlefield, 2016). Previously, he worked at Lockheed Martin, where he functioned as a systems engineer. He completed his MA in security studies from Georgetown University's School of Foreign Service and earned a BS in aerospace engineering from Pennsylvania State University.

Peter A. Garretson, Lt. Col., USAF (Ret.), is a senior fellow in defense studies at the American Foreign Policy Council, where he codirects the organization's Space Policy Initiative. He coauthored the well-reviewed *Scramble for the Skies: The Great Power Competition to Control the Resources of Outer Space* (Lexington Books, 2020). Prior to joining AFPC, Col. Garretson spent over a decade as a transformational strategist for the Department of the Air Force, where he served as a strategy and policy adviser for the chief of staff of the air force, as division chief of irregular warfare strategy plans and policy, and as chief of the Future Technology Branch of Air Force Strategic Planning.

Lamont Colucci is a senior fellow in national security affairs at the American Foreign Policy Council. A full professor of politics and government at Concordia University, Dr. Colucci has experience as a professional diplomat at the U.S. Department of State. He is an expert on grand strategy and great power competition.

Anthony Imperato is a research associate at the American Foreign Policy Council. Previously, he held positions with the Heritage Foundation's Center for Technology Policy and the Office of the Vice President at the White House. His publications have appeared in *The Daily Signal*, *RealClear Defense*, and *The National Interest*, among others.

Cody Retherford is a junior fellow at the American Foreign Policy Council. He has a background in counterterrorism, national security, and emerging technology research and analysis. He formerly served in the U.S. Army on active duty for five years in light infantry and special operations organizations. Cody holds a BA in international relations from the University of North Georgia and an MA in international relations and strategic studies from Johns Hopkins School of Advanced International Studies.

Larry M. Wortzel, Colonel, U.S. Army (Ret.), is a senior fellow in Asian security at the American Foreign Policy Council. Dr. Wortzel was appointed to the U.S.-China Economic and Security Review Commission for a term ending December 31, 2020. He has written and testified on space policy before Congress. He also has a strong command of the Mandarin language and understanding of China's global ambitions in the space sector and the resulting impact on U.S. military and economy.

Index